W9-CEW-531

LOVE UNDERSTOOD

Also available in the Bloomsbury Sigma series:

P53: The Gene that Cracked the Cancer Code by Sue Armstrong
Spirals in Time by Helen Scales
A is for Arsenic by Kathryn Harkup
Breaking the Chains of Gravity by Amy Shira Teitel
Herding Hemingway's Cats by Kat Arney
Death on Earth by Jules Howard
The Tyrannosaur Chronicles by David Hone
Soccermatics by David Sumpter
Big Data by Timandra Harkness
Goldilocks and the Water Bears by Louisa Preston
Science and the City by Laurie Winkless
Bring Back the King by Helen Pilcher
Furry Logic by Matin Durrani and Liz Kalaugher
Built on Bones by Brenna Hassett
My European Family by Karin Bojs
4th Rock from the Sun by Nicky Jenner
Patient H69 by Vanessa Potter
Catching Breath by Kathryn Lougheed
PIG/PORK by Pía Spry-Marqués
The Planet Factory by Elizabeth Tasker
Wonders Beyond Numbers by Johnny Ball
Immune by Catherine Carver
I, Mammal by Liam Drew
Reinventing the Wheel by Bronwen and Francis Percival
Making the Monster by Kathryn Harkup
Best Before by Nicola Temple
Catching Stardust by Natalie Starkey
Seeds of Science by Mark Lynas
Outnumbered by David Sumpter
Eye of the Shoal by Helen Scales
Nodding Off by Alice Gregory
The Science of Sin by Jack Lewis
The Edge of Memory by Patrick Nunn
Turned On by Kate Devlin
Borrowed Time by Sue Armstrong

LOVE UNDERSTOOD

The Science of Who, How and Why We Love

Laura Mucha

B L O O M S B U R Y   S I G M A
LONDON • OXFORD • NEW YORK • NEW DELHI • SYDNEY

BLOOMSBURY SIGMA
Bloomsbury Publishing Plc
50 Bedford Square, London, WC1B 3DP, UK

BLOOMSBURY, BLOOMSBURY SIGMA and the Bloomsbury Sigma logo are trademarks of Bloomsbury Publishing Plc

First published in the United Kingdom in 2019

A catalogue record for this book is available from the British Library

Library of Congress Cataloguing-in-Publication data has been applied for

ISBN: HB: 978-1-4729-6832-6; eBook: 978-1-4729-6831-9

2 4 6 8 10 9 7 5 3 1

Typeset by Deanta Global Publishing Services, Chennai, India
Printed and bound in Great Britain by CPI Group (UK) Ltd, Croydon CR0 4YY

Bloomsbury Sigma, Book Forty-two

Illustrations by Brian Southern

For Mum and Ed

And Stefan and Halina Greiner – loved and remembered

Contents

Introduction

I must have been an exhausting child as I was always, always asking questions. But there was one topic that fascinated me more than most – love and relationships.

That's probably because I didn't meet my dad until I was 29. I had little or no interest in contacting him for most of my life, but eventually decided that I might regret it if I didn't at least try. So, after a bit of (very basic) detective work, I managed to get hold of his phone number. Then, after two years of doing nothing, I rang him. He answered. It was the afternoon of his fourth wedding, and he and his new wife were watching the rugby.

Without my genetic father on the scene, I lived with my mum and spent most weekends with my grandparents. My grandfather (I called him 'Dad') was organised, patient and loving – at least with me. Apparently, he could be quite grumpy to others, but I can't really remember as he died when I was 11. I remember the phone ringing in the middle of the night – it echoed up the stairs and filled my ears with news of his death. We rushed to the hospice (not sure why – he wasn't going anywhere) to be greeted by his blue-grey corpse, theatrically draped in crisp white linen. His skin felt like cold wax. He was dead, and I was devastated.

I knew that losing him would have an impact on my life – I just didn't know how. But it wasn't just losing *him* that affected me. When he died, I lost the only committed romantic relationship I had been able to observe – that of my grandparents. As a result, I really struggled to understand what such a relationship could or should be. I had no idea what to expect and was left with countless questions: What is love? How does it change over time? Does love at first sight exist? What should you look for in a partner? How does your upbringing influence your relationships? Why do people cheat? What's the difference between lust and love? Should love be unconditional? How do you know when it's time to walk away?

From an early age – in fact, for as long as I can remember – I questioned, interrogated and informally 'interviewed' virtually everyone I met about their relationships. As I grew older and started travelling the world, I spoke to even more people from a wider variety of backgrounds, listening to their intimate stories and personal philosophies.

I must have asked millions of questions and spent years thinking about this before I met a farmer when I was travelling around Argentina 10 years ago. He was 95 and had been married for a whopping 75 years. He didn't speak very much, but I really valued the advice of a man who had managed to make a romantic relationship last such a huge amount of time. Then it occurred to me – I should document my investigations. I bought a recording device when I returned to England and started to carry it with me wherever I went.

Over the course of the next 10 years, I approached hundreds of people in airports, shops, markets, cafes, restaurants, bars, hospitals, parks, galleries, libraries, museums, buses, trains, planes and ships. I interviewed a pro American football player on a bus, a model who sat next to me on a plane, the bin man* who worked at the local train station, people who were religious, atheist, agnostic, male, female, transgender, homosexual, heterosexual, bisexual, single, married, separated, divorced, widowed, with children, without children, pregnant, monogamous, non-monogamous, cheating, cheated on, entirely faithful. I interviewed people in French, Polish, Spanish and English. I travelled over a quarter of a million miles, interviewing people from eight to 95 years of age on every continent of the world.

I randomly approached people and asked if I could interview them, explaining that it would take as much

* Or 'garbage collector' if you're reading this in the US.

time as they wanted (interviews ranged from four minutes to three hours). In most cases I recorded the interviews, although some people preferred not to be recorded. In these instances, I spoke to them while manically scribbling, then promptly wrote up my notes before I was unable to decipher my own handwriting (a trick I learnt early on as a lawyer).

I promised to preserve the interviewees' anonymity and change specifics that might identify them, reassuring them that if they didn't want to answer a particular question they didn't have to and if they wanted to retract anything, they could. I tried to make it clear that these interviews weren't about catching people out or showing them in a bad light – they were about trying to understand love, and to share knowledge and experience with others across the world.

Trust, non-judgement and confidentiality were crucial – without them, I wouldn't have received people's honest accounts, particularly with trickier topics like infidelity. Unsurprisingly, studies have found that people are reluctant to confess to cheating – one found that only 30 per cent of people initially admitted to infidelity, but during intensive therapy, a further 30 per cent confessed.

Recording, transcribing and editing the interviews gave me the time and space to digest and reflect on what I was hearing. And the more I thought about it, the more questions I had. So, I buried myself in academic textbooks and journals – highlighting, tabbing and writing copious notes. I'm not the only person to have been flummoxed by romantic love, but I may be one of the few with enough curiosity and stamina to *really* scrutinise it.

Initially, I focused on psychology and philosophy (which I studied at university), but I was soon delving into other subjects because, in order to really understand love, I wanted to attack it from all angles. Finally, my research came to life as I started talking to and questioning the academics whose work I had been devouring.*

The research was vital in understanding love because what we think or say isn't always the truth – and, what's worse, we may have no idea that what we are thinking or saying may not be true. As humans, we often try to find reasons for our behaviour. If we don't find someone attractive, we might blame it on their nose when it could actually be down to their pheromones or our attachment style. Academic theory and research (particularly neuroscientific studies or those involving physiological measures, such as heart rate, sweating, etc.) were necessary to explore the role of our subconscious, emotions and physiology.

Research also engaged my rational self – theories helped me think about things differently, make sense of my behaviour and thought patterns. Attachment theory, for example, helped explain my tendency to break up with people. But logic alone isn't always enough to create change and understanding. As Tristan in Switzerland said: 'Learning about attachment theory has been useful for me. The downside is that, while I can talk about things rationally, I won't necessarily understand them emotionally.'

That's where the interviews came in. Listening to people's stories forced me to pay attention by engaging my emotions. I really cared about the people I spoke to and their stories have often stayed with me. For example, Sue, who I met in Iceland, said: 'I was held against my will between the ages of one and nine, and incredibly bad things happened … During that time, I didn't know that love existed. When you're young, you think whatever is, is how it is. Then my captor enticed another child. I was

---

* A full bibliography of my research is available online at lauramucha.com/resources.

almost completely numb and mute at this point, but meeting this other child and seeing that this might happen to them brought me out of it and into the world … Strangely, from having a connection with somebody else, I realised what was important in life: connections with other people.'

I cannot overstate how much this project has changed me. It's impacted every relationship I have. I understand people differently, I listen more compassionately and I am infinitely less judgemental. And in terms of romantic love, I'm married – to someone I had previously broken up with. I still have lots to learn (marrying someone is one thing but putting in the hard work to make it last is another), but it really is an extraordinary feat given quite how commitment-phobic I was at the start of this process. Writing this book has made me question, consider, explore and reflect on my own life, and I hope it will help you do the same.

I really debated about how best to arrange the interviews and eventually decided to organise them thematically. I interviewed an equal number of men and women,* but you may find that gender is not equally dispersed throughout the book.

All that's left to say is how thankful I am to the interviewees and academics who were hugely generous in sharing their stories and philosophies. I ask that you read their interviews non-judgementally, with openness, kindness and, perhaps, a bit of love.

LM 2019

* I also interviewed people who didn't identify as men or women, including transgender people (who asked not to be identified as such in the book). My intention in referring to men and women is not to exclude anyone.

CHAPTER ONE

Don't Give Me Drugs

I was on holiday in Ibiza, staying at a villa that belonged to a charming couple who had popped by a few times to see how we were. Eventually, I asked the wife whether I could interview her and she obliged. We sat on opposite sun loungers overlooking a landscaped garden and an aquamarine pool.

I often change people's names to preserve their anonymity and, before starting the interview, I suggested that we call her Melissa.

'Oh, God, no,' she said, 'any name but Melissa … That was the name of the girl my husband had an affair with …' She let out a deep laugh and suggested another name, Lisa, before adding: 'In my life, love has had to be very forgiving, understanding, tolerant.'

Like many of the older people I spoke to, Lisa explained that love is much more exciting and heart wrenching when you're young. She told me that the youthful, more sexual side of love, while important, wasn't the be all and end all. Instead, it eventually turned to friendship love. And, she added, the older you get, the more you take it for granted.

She paused.

'I experienced an enormity of jealousy at one stage of my life, which was of three and a half years' duration. It all started when my husband went to the birthday party of a friend, and my husband thought that one of the guests was very young and shapely. She was 30 years younger than him. She started to talk to him about golf. She told him that she was dying to learn and he said, "I'll give you some lessons if you like." So he started to give her lessons – but not in golf.'

I can never, ever predict what I'm going to hear when I approach someone – nor do I try to. And I certainly didn't expect to be having this conversation with Lisa.

I began by asking how she knew her husband was having an affair. In subsequent interviews where infidelity* came up (of which there were many) I always asked this question. I'm not sure whether I was trying to gather information for the purposes of this project – or to protect myself. Perhaps both. But Lisa, like the majority of people I spoke to who'd been cheated on, knew before her partner confessed. Although partners didn't always confess, of course.

'I guessed something was wrong, simply because my husband started doing things that he didn't often do. He was constantly charging his mobile, always on the phone. There was a change of pattern in the time that he was coming home. The afternoon break that we usually spent together became a break where he would go off without me. He would return at seven in the evening and say, "I went to the golf club."

'He started to give me expensive presents. We went to the garage to get my car serviced and he promptly bought me a new one. The following year, he bought me another one. I read in a book somewhere that one of the first signs of trouble is when a husband gives expensive gifts to his wife.'

Lisa wasn't alone in noticing changes in what had been quite routine behaviour. For most of the people I spoke to,

---

*You might be wondering why I've chosen to start with infidelity – and it's a fair question. I think discussing infidelity is a great way of starting to explore the different types of love, as well as the distinction between love and lust.

that was the telltale sign – although the specific changes were different for each person. Marie, from Portugal, told me that her husband stopped going out with her, always using the same excuse: 'Don't worry, you go out without me – I think I'll just stay in.'* Even cheaters themselves recognised a change in their behaviour. Jessie, from Hong Kong, explained: 'Part of me thinks he knows that I have cheated. Obviously he won't know everything and every time, but he probably had a suspicion from me going out with girlfriends, coming home really late and hiding my phone so that he couldn't see the text messages that I was getting.'

For a few, though, there was no change – just warning signs along the way. John, from Australia, explained: 'She kissed people on the lips in front of me. She was very close to everyone, very affectionate, and of course that would often lead somewhere. It didn't come as a surprise when I found out that she had been cheating.'

Lisa had known her husband for a long time (they met when she was 14) and he'd had a number of affairs – although nothing 'of any duration or seriousness'. But this time was different.

'I could tell that this affair was vitally important to him by the way that he would quite literally run to get his phone when it made a sound. I eventually found out that, not only had it been going on for three and a half years, but it had been very serious. He had taken her to Barbados on holiday, Portugal, Spain, and all in the name of golf.

'It was heartbreaking to find out. It hurt immensely. It took me an awful long time to get rid of the evidence, all the letters that she sent him. And it took an awful long time

---

*You'll meet Marie and Jessie again in Chapter 9.

to believe anything he said ever again. I did eventually forgive and trust him again, but with difficulty.

'I didn't realise how seriously it had affected me. I always had a knot in my tummy. Eventually I developed a tremendous arrhythmia and lost three stone. My doctor, who knew the story, referred me to a consultant. Unbeknownst to me, she told the consultant everything and when I had my appointment, he said to me, "Your doctor tells me that you have suffered a broken heart – when your heart mends, all your symptoms will go." I thought broken hearts were things of romance novels.'*

I asked Lisa whether she had thought about ending the marriage. She told me that she hadn't and gave me three reasons for this decision: they had five children, they were business partners and she would hate to be lonely. Then she sat up and her voice changed, adopting a firmer tone.

'I will say that I do know that at no time, at no time, did he stop loving me. And at no time did he ever contemplate leaving me. I would consult lots of his friends and they would say, "He'll never leave you." I would reply, "Well, I might leave him..." But I didn't mean it. He is very loving to me. This morning, as I walked into the bedroom from the bathroom, he gave me a hug. Yesterday he said, as he walked through the front door, "I know what you're going to say – and I haven't had a drink – and I do love you very much." And I said, "You always say that when you've had a drink." He always says lovely things to me, like, "You're a very nice person."'

---

* They're not ... Stress cardiomyopathy (or 'broken heart syndrome') is a recognised medical condition that can happen to people who suffer massive emotional stress. The heart muscle suddenly weakens – it's essentially stunned – and the left ventricle (one of the heart's main chambers) changes shape. Apparently, it feels similar to a heart attack as the main symptoms are chest pain and breathlessness.

I asked her whether she had any advice for someone who suspected that their partner was cheating on them and she thought for a moment. 'I would first establish whether the cheater was a partner or a spouse. If it was a partner who cheated, I would break away from it. Hard as it may be, if they have been cheating, and have been doing it for some time, I would say that the cheating is going to cause you an awful lot of heartbreak, so it might cause you less heartbreak to break away from it. You might say to them, "I know what's going on, I am still very fond of you, but I'm not prepared to be cheated on. Unless you're prepared to make this relationship work and give the other person up, I would suggest that the best thing we can do is go our own ways." If they love you, they will end the other relationship and come back to you.

'Once you are married, I think you have to be more tolerant of misdemeanours. You would be very lucky these days to find a marriage that doesn't involve big dips and falls and hasn't had indiscretions. Firstly, contraception is easy these days – it is much easier to have flings and go the whole way. When I was a girl you had to be very careful, whereas now you don't have to be. And even if you haven't been careful, you can take the morning-after pill. Secondly, technology means that you can discreetly keep in touch in a number of ways – with phones and texts and social media. It's much easier to hide your affairs now.'

The sun began to set and the aquamarine pool had turned purple-navy. We had both noticed the drop in air temperature and Lisa said she really should be getting home. She wished me all the best with the project before leaving. I thanked her for her honesty and watched as she walked off through the manicured trees.

And that was that. Except my conversation with Lisa stayed with me for a long time (like many interviews, I doubt it will ever leave) and for months, the idea that most people cheat rattled around my brain like a ghost in a locker room. It also confused me. In a study of 2,025 adults in the US, 88 per cent* believed that it was totally unacceptable for a married person to have an affair. If most people hated infidelity, surely they wouldn't *also* be cheating?

I scoured research papers from all over the world, analysing details such as age of participant, length of relationship and gender – and, to my horror and surprise, discovered that Lisa was probably right. Infidelity is in fact very popular – but just how popular is difficult to tell.

A prolific pastime

The problem in studying infidelity is that most people don't want to admit to their indiscretions (presumably for fear of being found out, but perhaps also for fear of being judged). Maybe this goes some way to explaining why studies have found a huge range of results: depending on the study, 14–72 per cent of men and 10–70 per cent of women admit to infidelity.

The eighteenth-century philosopher David Hume did not have a particularly high opinion of women's fidelity, writing, 'What restraint … shall we impose on women, in order to counter-balance so strong a temptation as they have to infidelity?' I have to give it to Hume as there is some truth in what he says: women, whether in gay or straight relationships, do cheat. But most studies suggest that men cheat more. Well, either that, or they're more willing to admit to it.

It's hard to get to the bottom of just how much more men cheat as it varies wildly. In one US study, 25 per cent of

* Unless there is good reason not to, I'm going to round up or down to the nearest whole number to try to keep things simple.

men admitted to it, compared with 15 per cent of women. But in another, the difference was much smaller, with 26 per cent of men and 21 per cent of women confessing.* Meanwhile, an analysis of numerous studies looking at more than 90,000 people across different countries found that more than a third of men cheated, compared with less than a quarter of women.

I should pause briefly here to point something out. Infidelity is, unhelpfully, defined differently depending on the study – some define it as kissing, others sex and some a full-blown affair. (Most do not include emotional infidelity, which I'll come on to later in the book.) This analysis of 90,000 people compared 50 different studies, many of which had different definitions of what it is to cheat. And the lack of a universal definition helps explain why different studies have dramatically different results.

In any case, most studies find that, despite cultural disapproval, cheating is still a relatively popular pastime. One of the more public confirmations of this was the Ashley Madison leak. The online dating service, marketed to people who are in committed relationships, was hacked in 2015 and the data from over 36 million users was leaked. The hack proved that the site was incredibly popular – 6 per cent of Canada's entire population (over 2 million users) and 0.05 per cent of the US (16 million users) had signed up.

Just in case that didn't fully register, that's 18 million users in the US and Canada – almost four times the entire population of New Zealand. It was particularly popular with men, who accounted for 86 per cent of users, although this may say more about how the site was marketed and the

---

* In this second study, men cheated with more people than women. And the numbers were higher for couples who lived together but weren't married, with 33 per cent of men and 30 per cent of women admitting to infidelity.

way people go about being unfaithful than the gender difference in doing it.

You may think that the statistics I've cited so far are astonishing, but the percentage of people cheating is likely to be much higher. There are plenty of people who are unfaithful and have not used Ashley Madison. And in a study of 7,239 men, 66 per cent of those who were married admitted to sleeping with someone other than their spouse (almost three times the 25 per cent cited in the study I mentioned earlier).*

This is an important study because, as well as being relatively large, the sample was representative of the US, which means that the characteristics of the people involved in the study match those of the US population.† Whenever academics want to study a small number of people and generalise their findings to a larger one, the ideal is to make the two groups as similar as possible. But it's hard work – it's much easier to ask undergraduates to take part in studies as a requirement of their course (which is what happens in a massive amount of research).‡ Because this study was representative of the US, it provides a particularly good

---

* I should add that in 3 per cent of these cases, the spouses knew about it or they had an open marriage. The percentage went up to 72 per cent for those who had been married for two years or more. I haven't provided the rates of infidelity in women in this study because the researchers didn't explore that.

† I should mention that this study is relatively old, taking place in 1981. But the sample was larger and more representative than most of the more recent studies.

‡ One more thing … Even when a sample *is* representative of, say, the US population, those findings can only extend to the US and not to other countries, like Bhutan or Botswana. And this is a problem because 96 per cent of samples in psychological research come from Westernised countries that make up only 12 per cent of the world's population.

insight into how often infidelity is likely to be happening (at least in that country).

Research into LGBTQ+* relationships is thin on the ground, but what there is suggests that, as in heterosexual relationships, men cheat more than women. A study of 776 lesbians found that 28 per cent had slept with someone else while being in a relationship. Whereas a study of 242 gay men found that 45 per cent of those in a monogamous relationship had slept with someone else. Frequency of the rule-breaking sex ranged massively from two to 2,500 incidents.

Given that most people are reluctant to admit their infidelity, these numbers should be taken as a minimum. They should also be seen as conservative because someone may not have cheated by the time they were questioned by researchers, but might go on to later in life. What we do know is that people – both men and women – cheat, even if they – and their partner – are vehemently against it.

Despite my initial shock at Lisa's statement that I 'would be very lucky these days to find a marriage that … hasn't had indiscretions', the numbers suggest that she was right. But whether or not she was right to stay with her partner is another question: a question that, if the data is correct, many of us may have to face.

* LGBTQ+ stands for lesbian, gay, bisexual, transgender (people whose gender identity or expression differ from the sex they were assigned at birth) and either queer or questioning (depending on who you ask). (Some people use LGBTQIA+ which also includes intersex and either ally or asexual.)

Don't give me drugs, give me a married man

I was often surprised by – and thankful for – the trust, confidence and honesty that interviewees gave me. My aim has always been to better understand love and romantic relationships, and I was only going to achieve this if my conversations were honest and personal.

So I was very excited when I started talking to Claire in a cafe in Frankfurt.* She was in her 60s, wore thin-rimmed reading glasses and was sitting alone, reading a newspaper. She lived in Germany but was originally from the US, and the first thing she told me was that over the course of her 20s and 30s, she dated three different married men (although not at the same time).

'It all stemmed from insecurity. The only reason I went out with them is that they gave me attention and they liked me. I didn't particularly fancy people. Instead I was drawn to people that were drawn to me – that way, I could finish it whenever because I hadn't chosen them. I had such low self-esteem that I went with these married men, not thinking about their wives or their families. I would just put that to the back of my mind; I blanked it out completely. When I think about it, I didn't know who I was or what I wanted. I was just fumbling around.'

Claire clutched her coffee with both hands, deep in thought before describing the restrictions that these extra-marital relationships faced, some of which would never have crossed my mind.

'I couldn't buy for them and they couldn't buy for me. If I gave them a present they would have to go home and lie about where they got it from – and that would denigrate what I had bought. I didn't have a domestic life with them,

* Just to warn you, this chapter features more women than men, but I redress this when I revisit the subject in Chapter 9.

we didn't cook for each other, we rarely stayed overnight; in retrospect, the time together, as well as the relationship itself, was very limited, so I wasn't getting a true picture of who they were.

'When we were together, we were always watching over our shoulders to make sure that we didn't see anybody we knew. It was exciting, but on the other hand, we were living a dangerous existence. In one relationship, we had to go outside the area that we lived in because staying there was too risky as we might be seen by someone who might report back. It took a lot of energy and it was quite stressful.'

The excitement and danger that Claire talked about reminded me of a famous experiment that explored the impact of circumstance (and, in particular, fear) on how we interpret human interactions. An attractive female experimenter stopped men aged 18–35 and asked them to complete a questionnaire (she only stopped men who weren't accompanied by a female). When they had finished, she tore off a corner of the page, wrote down her name and number, and asked them to give her a call if they wanted to talk further.

Men were stopped on one of two bridges: the first was made of wooden boards attached to wire cables, which had a tendency to tilt, sway and wobble. It was 137m (450ft) long, had very low handrails and a 70m (230ft) drop to rocks and rapids below. The second was a wider and firmer bridge made of solid wood. It didn't tilt or sway, had high handrails and was 3m (10ft) above shallow water. Now here's the interesting part: men who were stopped on the wobbly bridge were more than four times as likely to call the experimenter than men stopped on the secure bridge.

These findings suggest that fear can make someone seem more attractive than they actually are, perhaps because we attribute increased heart rate to meeting them rather than the fight or flight response caused by a wobbly bridge.* (Of course, men who chose to cross a wobbly bridge may have been more likely to be risk takers and therefore more willing to risk phoning a female they didn't know …)

Perhaps the danger, fear and excitement that Claire felt when meeting the married men made them, and the relationship, feel all the more intense – when it was actually just the fight or flight response kicking into action because they were both worried about being caught. I wonder what they would have felt about each other had they met in the post office queue on a rainy day when they were both single.†

I asked Claire whether she loved any of them and she paused.

'There was obviously something there between us because otherwise it would not have continued. But … I'm not sure whether I loved them – although I thought I did at the time. To be honest, I don't know whether I really knew them very well because I only saw one side of them. The whole thing was built on a lie and when something is built on a lie, it has no foundation.'

She wasn't the only person to comment on the impact of building a relationship on dishonest foundations. Ethan from Wales told me: 'Many of my relationships have started via me stealing someone or being the accomplice to cheating. In most cases, the infidelity ended their relationship and I ended up with this new woman. This is what happened with my ex – I stole her from another guy.

---

* So, if you're looking for love, you could do worse than lurk, casually, on wobbly bridges, cliff edges or at theme parks in the hope that this will make you appear more alluring.

† I explore the impact of external factors like music, crowding, temperature and alcohol in Chapter 5.

She then went on to cheat on me with my best friend. I don't think I've seen the irony until now that I find it hard to trust the women that I steal, because that's how I ended up being with them – through their lying to their partner at the time.'

Claire put down her coffee, lent back in her chair and started drumming her fingers on the armrest. 'I think they were a bit bored with their lives. They had been married a while, they had kids. They each made excuses for their cheating, like, "My wife doesn't understand me," which meant that I could say, "Don't worry, I understand you." A part of me did want them to leave their wives, but in the back of my mind, I probably didn't want commitment either.* I knew that if they left their wives, that would have thrown our relationship in a completely different direction – and it probably would have ended.

'I just enjoyed it for what it was. There was a certain thrill – when we spent time together we just enjoyed it, there was something exciting about it. We were always left wanting more. I got all the thrills but without the responsibility or commitment. I didn't see them that much, so it didn't impinge on my life. When I wasn't with them, I was sort of single again, in a mirage of a world. It was quite selfish. In a way, it was like a drug to me. Don't give me cocaine, don't give me heroin, give me a married man.'

When Claire talked about the drug-like high that she experienced with these married men, it got me thinking about the huge 'highs' I had experienced when meeting certain people and just how powerful they could be. And as I went on to speak to more and more strangers, the comparison between lust (or love) and drugs came up

* Going for someone who's unavailable is a classic strategy for people with an 'avoidant attachment style' – I'll come to this in Chapter 3.

again and again. It left me wondering what was going on in the brain to cause something so potent (and often untrustworthy).

In 1954, James Olds and Peter Milner implanted electrodes in the brains of rats and allowed the animals to press a lever to receive electrical stimulation to their brains. Olds and Milner soon realised that when it caused electrical stimulation to certain parts of the brain, rats would press the lever over and over again – with one of the rats pressing a lever 7,500 times in 12 hours.

Their experiments confirmed that lurking deep within the brain were structures devoted to pleasure and reward – and further studies confirmed just how powerful these structures were. In 1965, Aryeh Routtenberg and Janet Lindy found that when offered a choice of food and water *or* electrical brain stimulation, rats opted for starvation in order to get that high.

We now know much more about the structures in the brain responsible for pleasure and reward, which are collectively known as the brain's 'reward system'. When it's activated, dopamine passes from one area to another* and our brain tells us to repeat whatever we just did to get that reward. Dopamine is essentially a messenger – a chemical that transmits signals from neurons (nerve cells in the brain) to other neurons. It doesn't know, or care, what or who caused the high; it only 'knows' that it wants the high again.

Female prairie voles who were raised in labs started to prefer one male above all others at about the same time they experienced a 50 per cent increase of dopamine. But when prairie voles were injected with something to restrict dopamine activity (a dopamine 'antagonist'), they preferred

* If you want to know more: dopamine passes from the VTA (or 'ventral tegmental area') to something that looks like a little nugget, the NAc (or 'nucleus accumbens').

the vole that happened to be around at the time, even if they hadn't mated with him before. Similarly, rats that pressed the lever over and over and over again for electrical brain stimulation lost interest when injected with a dopamine antagonist.

Research in humans has shown that we also find electrical brain stimulation in this pathway deeply pleasurable – so pleasurable that participants likened it to intense sexual orgasm, with some developing a strong romantic attraction to the researchers performing the experiments.

For ethical reasons, human research in this area has been limited* – but another study suggests that the reward system is activated when we are 'in love'. Psychologist Arthur Aron and colleagues put people in an fMRI scanner and showed them a photograph of a neutral person and the person they had 'fallen in love with' a few months earlier. When people looked at the photo of their beloved, the brain's reward system was the most active part of their brain,† and it's the brain's reward system that is activated if we take drugs.

Cocaine and amphetamines, for example, increase dopamine in the reward pathway, which is perhaps why love or lust can feel like the high you might get from using narcotics. In fact, love or lust might be even more potent. Sarah Leibowitz, a specialist in how the brain behaves when it comes to addiction, believes that while drugs can induce powerful changes in the brain, none do so as reliably or delightfully as that 'right' person.

Something else struck me about the drug-like high that Claire said she experienced with these married men: it felt

---

* Pesky ethics getting in the way of juicy experiments …

† Of course, loving in real life is very different to looking at a photograph of someone you love. Love is not simply a response to a stimulus and scanning one area of the brain won't capture the complexity of what is going on – but it can give us an idea.

completely different to what is likely to have been a far less exciting love that these men may (or may not) have experienced for their wives. This is one of the reasons I decided to begin the book by exploring infidelity – it's a fantastic way to highlight the differences between love and lust, as well as the distinct types of love.

I will explore this more in the next chapter, but for the moment, I want to make a very basic point: there are different types of love, they feel different and their underlying neurological basis is different. One type – the exciting, passionate type – engages the brain's reward system. This love is very powerful – and deliberately so, because without it, we probably wouldn't survive as a species.

Another type, the calmer, much less heated friendship love that Lisa mentioned, involves a different set of neurochemicals and isn't as dominating. Instead, like friendship, this sort of love is more about companionship, respect, vulnerability and trust, as well as developing a deep knowledge of each other. This sort of love might be more concerned with sharing values than bodily fluids.

When people think and talk about infidelity, they are often comparing two very different types of love (or they may be comparing love with lust ...). It's a bit like comparing a glamorous racehorse with a trusted donkey. They're not the same thing. What's worse is that the glamorous racehorse is made even more alluring by the wobbly bridge effect – or in this case, the danger of getting caught. Given what it has to compete against, perhaps it's unsurprising how often the trusted donkey loses out.

The cafe door swung open and a rush of cool air hit my face. Claire readjusted her glasses, which had slipped to the end of her nose, before telling me about her current relationship. After some therapy, she eventually stopped dating married men and finally married a kind, loving (and unmarried) man. She admitted that her feelings for him were not as passionate as those she'd had for the married men, but, after 12 years together, she could say without doubt that she loved him. She told me that, finally, she was happy – and she seemed it.

I asked her what, if anything, she would say to her younger self. She stared out of the large coffee shop window and said: 'I was very naïve, very limited in my thinking. Thinking about it now, I find it very disrespectful. I have had a complete round turn of my view on it. If I could speak to my younger self I would say, "Don't do it – he's already been very selfish in going for you, he's being disrespectful to his wife, he's lying, he's cheating." To a certain degree, it's greed on his part. Even if he says, "I'm going to leave my wife," what are the chances that he would do the same thing to you? I would tell my younger self to say to him, "Sort yourself out and if I'm still available, then we'll take it from there. But right now, you'll actually be adding to your problems if we get together."

'There's something wrong with either you or your relationship for you to want to cheat, to want to hurt your partner – because that's what you're doing. You're cheating on them. They don't know it, but you know it, so it's on your conscience. It's not fair to anyone involved. Trust is such a major and precious thing. Once you cheat, you damage the relationship irreparably because it's like a scar. It will always be there, you will never get rid of it.

'His priority, especially if there are kids involved, will be to the wife and kids, so if you are the girlfriend, you will always come second. I think it's a losing game, even if the end result is that he leaves his wife and goes for you. Most of the time, of course, he won't, as he just won't want to make a decision.

'I would say to my younger self, "Look at what it is in you that is wanting this." Everything stems from morals and kindness. If you were being kind to yourself and to other people, you wouldn't do these things. I wouldn't have done these things.

'But on the other hand, sometimes I wonder whether we do need to go through these things in a perverse way to develop ourselves. How else do you become who you are?'

CHAPTER TWO

A Promiscuous Word

Infidelity is a hefty topic, and one that I will return to later in the book, but I wanted to pause and make sure we are clear on what exactly we're talking about. And that's not easy because love is a promiscuous word. Oli, a librarian in his 50s,* described it brilliantly when he said: 'There are so many varieties and variations of love, it's almost impossible to talk about it. Are we talking about familial love? Sexual love? Whether I love my cat? Whether my cat loves me?'

The ambiguity of the word makes it very difficult to explore. You might think that talking specifically about romantic love would solve the problem, but even that's used to describe a variety of things – some of which you may not think are 'love' at all.

The Ancient Greeks got around this by having seven different words for love (six of them Greek, one Latin),† which makes sense in light of its myriad meanings – but English, which has official or special status in over 75 countries with a combined population of over 2 billion, has

---

* Oli overheard me doing an interview with someone else and came over to share his views on the subject.

† In case you're wondering, they were: *storgē* (natural affection, the sort of thing parents feel for their children); *érōs* (romantic love, meaning passion and physical attraction); *ludus* (the playful affection between children or casual lovers – this is the Latin word); *philía* – friendship (literally translated as 'love'); *pragma* (mature love, including patience, compromise and tolerance); *philautia* (self-love, which can either help us to love others, or be negative and selfish); and *agápe* (selfless, unconditional love that expects nothing in return).

only one word for love. I often wonder whether, if we had more words for the different types, we might appreciate and consider the less glamorous forms a little more – or at the very least, there would be less scope for confusion.

Defined terms

In a legal contract, you'll see some defined terms – a sort of mini-dictionary aimed at making sure everyone knows what they're talking about and agreeing to. That way, everyone is (or should be) on the same page.

One of the frustrating things in researching love is that there are no agreed terms used by all types of love researcher. Or rather, there are some agreed terms here, others there, and some who do studies about 'love' without defining anything at all. This means studies that rely on people saying they're 'in love' may actually be measuring something else. They might, in fact, be measuring lust, particularly in the early stages of romantic love when the two often overlap.

None of this is helped by the fact that there are more than 30 different measures of romantic love used by psychologists – as with infidelity, different studies may compare different things, depending on which definition or scale is used.

So, I thought that, at the very least, you and I could work from the same definitions (which I will refer to throughout the rest of the book):

Lust (passion, sexual desire or the sex drive) is a heady, intoxicating, obsessive and powerful drive aimed at getting us to reproduce. It can be directed at different people, lasting only moments – but can also be directed at (mostly) one person, lasting months or years. That said, it's not necessarily consistent – it can come and go in waves, influenced by our hormones* and behaviour.† Lust can be negative, using others to satisfy our own ends, but can also be reciprocal, fulfilling and meaningful – without being love. Lust is usually a feature of romantic love (which can make it difficult to tell them apart).

Romantic love (being in love, falling in love, passionate or erotic love) is made up of intimacy, which requires time and vulnerability, and lust – but not necessarily commitment (which I'd argue is necessary for a successful long-term relationship). Like lust, it can involve a certain amount of idealisation at the start and, the vast majority of the time, it dwindles – but not always.

Companionate love is a calm, stable love, often seen in mature long-term romantic relationships. In companionate love, both partners are honest and (relatively) realistic about both the relationship and each other's complexities and contradictions. Companionate love looks and feels much more like committed best friendship than a heady, passionate infatuation. Unlike romantic love, lust may not be there from the start, it might come and go, and it might

---

* To give you a flavour of the research, when men sniffed a T-shirt that had been worn by either a woman in the few days before ovulation (when she was highly fertile), or a woman much later on in her cycle (when she wasn't), researchers found that with one whiff of fertility, men's bodies started to gear them up for desire by raising their testosterone levels – even though they had no idea this was happening.

† For example, research suggests that there are things you can do if you're in a relationship to boost sexual desire, including making your partner feel desired and valued, trying to meet their sexual needs without expecting an immediate return, doing new and exciting activities with them, having positive, supportive and intimate conversations with other couples and believing that you can be sexually satisfied in your relationship through hard work and effort.

be less intense or frequent. Instead, partners may develop a deeper but more relaxed attraction to each other.

I'm not quite sure where 'chemistry' fits in – it's probably partly lust, partly some of the biological processes that I describe in later chapters. What I can say is that I have a deep suspicion that people overvalue chemistry and undervalue commitment when looking for a long-term partner. But I'll explore that more throughout the book.

I'm also unsure exactly what the 'honeymoon period' means – but in the interest of simplicity I'm going to equate it with lust, as this is how the interviewees mostly referred to it.

Right, now that we've got that out of the way, I'm going to explain how I reached these definitions.

Love versus lust

I've already described how studies have shown that rats opt for starvation to get a 'hit' from the brain's reward system – the part of the brain that's stimulated if we take cocaine or amphetamines. I also explained that this is the part of the brain that's activated if we meet the 'right' person.

Meeting someone can quite literally be like taking drugs. In fact, some think it's even more potent. But what exactly is behind all of this: lust, love or both?

Some interviewees were ambiguous when talking about love – using that one nebulous word to describe a plethora of experiences and ideas – but others were careful and specific, particularly when it came to the difference between love and lust (or 'passion').

One of my favourite explanations of the difference between lust and (companionate) love came from Terri. I met her in Denver airport in the US where she was sitting huddled up in a doorway, waiting for a flight home to Phoenix, Arizona.

'Are you kidding me? There is such a difference between lust and love and it's so hard to distinguish the two, and that can be dangerous itself. I think lust is just another form of love.* And something that is crazy about the word love itself is it's just one word and it encompasses so much … It's not right, there should be more words.

'I think lust is like a firework and love is more like a slow burn, it's a campfire burn, you just have to add tinder to it, you have to take care of it and look after it otherwise it goes out. But lust is definitely like BANG, BOOM and then once you get it out of your system that's it, it's over. But lust can lead to love, it can.'

Sean, a pro American football player I sat next to on a bus, was also very firm on the distinction, saying: 'Don't tell someone you love them when actually it is only lust. There is a fine line between the two.' And Michael, from Bristol, England, thought it was important not to confuse them, saying they 'are as different as walking and driving a car. The two can overlap – and hopefully they do for lots of people – but I don't think confusing them is terribly useful.'

The power of lust

Given how often I was warned about mistaking lust for love, one of the first things I asked philosopher Simon Blackburn when I† met him at Trinity College, Cambridge, was whether he knew of any way of telling them apart. I quickly realised it wasn't that simple. For a start, they can both be incredibly powerful, which

---

* I have to say I don't agree with her on this – I don't think lust is a type of love. I think you can feel lust at the same time as feeling love for someone, but you can also *just* feel lust (without any love).

† Technically I should say 'we' – I met him with my brilliant radio producer, Marie Horner, who recorded the interview.

doesn't exactly help with clear thinking. Blackburn explained that lust has 'the tendency to drive other things out of your mind and concentrate it on whatever you want. Of course, that makes it easier to criticise because the words "excessive" or "extreme" and so on already imply that you're out of control – that you're a victim of lust.'

Not only can it take over your mind and focus your thoughts on the other person's positive attributes, lust can add a few more for good measure. The French novelist Stendhal likened this to the process of 'crystallisation' seen in the salt mines of Salzburg. If you throw a stick into a mine and pull it out a few months later, you'll find it 'studded with a galaxy of scintillating diamonds', and the original branch won't be recognisable. He claimed it's the same when you fall for someone: you don't see them realistically, but instead encrust them with your unrealistic sparkling crystallisations.

Stendhal was actually talking about love when he described the process, but according to Blackburn, this crystallisation can happen in both love *and* lust. If that's true, it would explain why it can be so difficult to distinguish between the two. In fact, you may not know whether it's lust or love until the dazzle dissipates. 'I'm not sure you can tell the difference from the inside at the time,' said Blackburn. 'It's one of these things we don't know about ourselves sometimes. You learn about it pretty soon … But you may only know that after the event. I don't think there's any diagnostic you can be sure of.'

So, in those early, heady stages, you won't know whether you're dazzled by someone because they're dazzling or because you've mentally coated them in diamonds. You'll only know for sure once the sparkle subsides – and that takes time.

But time might not be enough. Getting to know someone – and allowing them to get to know you – also requires intimacy, and that means each person has to be able and willing to be honest about who they are, including the grisly and uncomfortable bits. In other words, they have to be vulnerable with each other, which isn't always easy. In fact, it can be very difficult and people use plenty of subconscious strategies to avoid it.

Lust, on the other hand, is much simpler – people can have a passionate relationship without any knowledge, (real) intimacy or vulnerability. But they may still be convinced, thanks to all the fizz, that what they're experiencing isn't lust but love.

Different types of lust?

One of the reasons lust can be so powerful and convincing is that, without it, we would die as a species. Most psychologists agree that its biological purpose is to get us to reproduce – this makes it potent, but also universal. It doesn't matter how clever, powerful or successful someone may be, no one is immune to the clutches of lust.

In light of its driving biological power, it's easy to see it as a bad thing. This was certainly how philosopher Immanuel Kant saw it (as far as we know, he never had sex). Kant saw lust as something that doesn't care what the other person thinks or feels, but instead objectifies, dehumanises and degrades them, treating them as nothing more than a means to satisfy their own ends. (It's not just lust that troubled Kant – he seemed to be disgusted by almost any kind of sex, including masturbation, which he considered a crime worse than murder.)

But perhaps lust doesn't have to be bad. Perhaps there are different types: Kant's, which is driven by pure passion, and another, which is reciprocal and involves more than just sexual pleasure.

This more mutual type of lust fits with the thinking of philosopher Thomas Hobbes. He thought that part of lust's enjoyment is the mental satisfaction we get from giving physical pleasure to someone else. Of course our own enjoyment plays a part, but this sort of lust is more of a conversation than a lecture.

Blackburn explained: 'Hobbes says this very nice thing about lust. He says that the exercise of lust is a physical pleasure, but it's not only that. It's the delight that you take in being able to please somebody so much. The fact that you're giving so much pleasure to somebody else is itself a cause of pleasure. And I think that's a great insight actually, and it is of course only true when things are going well, when you've got a reciprocal relationship and you're both into it.

'But it's an ideal which ought really to be better known because too much … of people's ideas about love, lust or sex tend to be ideas of doing it to somebody else or domination … And that implies one-way traffic – it doesn't matter what the other person thinks or is like or feels or whether they are into it at all …

'Hobbes' ideal is different – it is an ideal of reciprocity, taking pleasure in the other's pleasure, a sort of ping-pong ball. And I think that is very important and ought to be much, much more emphasised because lust without that is eventually going to be unsatisfying because the human dimension is not there. And unless you're a very, very basic human being, you'll want the human dimension.'

If Hobbes and Blackburn are right, lust can be meaningful in that it can create something mutual where each cares about the other in some way. If that's the case, lust doesn't have to be the ugly sibling of love, but can be enjoyed and celebrated for what it is.

But when it's mutual and meaningful, lust is all the more easily mistaken for love – and in light of its biological power and purpose, this confusion can have disastrous consequences. Elinor, an Irish lady I met in Paris, explained: 'The most important thing I've learnt about love is that lust is so different to "real" love … When I got married, I wasn't in love, I was in lust.' After two years of marriage, she realised she'd made a mistake; she divorced and developed multiple sclerosis shortly afterwards – which she attributed (in part) to the break-up.*

Goodbye passion, hello friendship

Having been a nun for 20 years, Katherine decided to leave the community and get married. When talking about her relationship, she said: 'There was a huge amount of romance initially. I remember when a friend came round and said to me, "You must go and have a look at your loo." I went in and saw that Jim had decorated the entire room with loo roll that said, "I love you" all over it. It was very embarrassing, I can tell you, but it was so sweet of him.

'As time goes on, the romance lessens and the friendship strengthens. The main thing is that we are close, we want to share things and we are committed. I'm now with Jim forever. He irritates me, I get angry, and all those sorts of things. I suppose each of us has to be a bit unselfish not to stick to our guns. But underneath, we're together and that's it. This is how it is.'

'At the beginning it is butterflies, excitement and a lot of movement in your body,' said Leo from Switzerland. 'I think that first excitement is a combination of lust and getting to know a person, figuring out if they are someone you could love. A Swiss-German might say "*Ha di gärn*" – it means "I like you, I like you a lot." Maybe you will go to "*Ich liebe dich,*" which means "I love you," later on. For a Swiss-German, "*Ich liebe dich*" is loaded with meaning and commitment, so it takes some time to actually utter it.

---

* I should add that we still don't know what causes multiple sclerosis (although certain factors, such as smoking, have been linked to it).

'Over time, it calms down. It's like if you have a sip of whisky and warmth spreads all around in your belly – it is a warm feeling, a feeling of familiarity, feeling drawn to someone. Lust doesn't play a major part. I think it is more a sense of connection, a sense of being together. I like to talk to her all the time, I want to talk to her and that is a part of love too for me.'

Lily, who I met in Dubai, told me that her mother had scolded her for being naïve when two friends called off their wedding. Lily thought they were breaking up because they didn't love each other any more, but her mother said: 'Well what the bloody hell has that got to do with anything? You youngsters think everything is about love. You do love somebody, of course you do, but over time, the love changes and you might find that you love your partner as a sibling. You all need to grow up a little bit.'

Romantic love

Psychologists have come up with plenty of theories about the different types of love, from whirlwind romance to love at first sight; from romantic love with obsession to romantic love without. But what most agree on is the difference between 'romantic' and 'companionate' love.

Romantic love (also called passionate or erotic love) is intense and leaves us feeling slightly anxious and desperate to be with that special person. And that's probably because it's linked to higher levels of dopamine,* noradrenaline and cortisol, and lower levels of serotonin.

Dopamine essentially tells us to go back and do whatever it is that we did in the first place to get that 'high' of the brain's reward system. In the context of romantic love, that usually means getting close to our 'loved' one.

* We met dopamine in the previous chapter and explored its role in pleasure and drug use.

Noradrenaline (also called norepinephrine) plays a role in anxiety and fight or flight – it mobilises the body for action and focuses attention.

Cortisol (also known as the 'stress hormone') plays a role in stress, fear and pain (which might help explain why romantic love can make us feel highly alert and cost so much energy).

Meanwhile, serotonin can reduce certain types of activity in the brain, such as repetitive thoughts. So experiencing a drop in serotonin levels can cause the intrusive thoughts, hopes and fears of romantic love. In fact, one study found strong similarities between the serotonin levels in brains of people who'd recently fallen in love and those with obsessive compulsive disorder (OCD).

Companionate love

Companionate love (which is similar to friendship or sibling love) is the tenderness we feel for someone whose life is entwined with ours. It's calmer and less dominating and can come with a much deeper attraction than with romantic love. It's linked to both vasopressin and oxytocin release – the former helps us form social and romantic bonds, and the latter (also known as the 'love' or 'cuddle' hormone) makes us feel calm, contented and secure.

Most philosophers agree that in order to be friends or companions, people must care about each other and do so for their own sake (not simply because the relationship is enjoyable or useful* – although it may be both). Most also agree that this means helping them without any ulterior motive, feeling joy in their success, and disappointment and frustration at their failures (rather than disappointment

---

* Aristotle classes these as two types of friends (pleasure friends and utility friends), but many contemporary philosophers don't see this as a friendship at all.

and frustration at them as people). On top of that, some philosophers argue that friendship requires a certain history because without it friends can't have a deep understanding of each other and what's important to them.

Compared to most of our friendships, companionate love is more committed and more intimate, but intimacy is a sticky subject, as philosophers disagree on precisely what it means.* Some say it's just a matter of sharing very personal information, but this underestimates the kind of trust involved, reducing it to keeping secrets. Others argue that it involves trust in someone's goodwill towards us (and not just in relation to the secrets shared), but also their judgement of what's in our best interests. In an ideal world, we'd be able to rely on them to intervene if we were doing something self-destructive, like neglecting an important someone or something.

Some argue intimacy is even more than this; they suggest that trusting someone's judgement means not only trusting that they understand who we are, but also what we find important in life and why. In other words, our friends need to understand, and perhaps also share, our values and what we find meaningful.

Goodbye passion

The two types of love are very different: romantic love leaves us feeling excited and anxious, whereas companionate love makes us feel calm and secure. And perhaps that's because they have different purposes – philosopher Arthur Schopenhauer saw romantic love, with all its ideals and illusions, as the result of our drive to reproduce. He thought its aim was to get two people obsessed with each other for long enough to produce and raise the next generation.

---

* I explore what commitment means in Chapter 8.

This was something that Nick, who I interrupted while he was skateboarding, also commented on: 'There might be biological preferences as to why you feel you love someone – it's not necessarily romantic to want to reproduce. You might think you love someone when actually it's part of a wider aspiration to create another version of yourself, quite narcissistically.'

But that's not the aim of companionate love. Some think it evolved so that we'd hang around long enough to look after any offspring and bring them up in a calm environment, but I often wonder whether its purpose is not reproduction but safety and security – perhaps we form these bonds because they help us survive, because, as humans, we're not wired to be totally independent.*

Most psychologists think that romantic love wanes, but they don't agree how long it sticks around for. Theories range from 30 years (in a study measuring dopamine activity), 29 months (in a study looking at brain activation patterns) and 12–18 months (in two studies investigating serotonin transporter levels in blood and hormonal changes).

Perhaps the huge range of results is down to just how complex romantic love is – and as a result, measuring cortisol or dopamine alone won't tell us the whole story. Or perhaps romantic love isn't consistent over time. Psychologists Brenda Munro and Gerald Adams found that in long-term relationships, romantic love started and ended well, but dipped in the middle. They suspected that romantic love took a nosedive if and when partners became parents, but went back up when the children left home.

Munro and Adams are, however, in the minority – as are those who argue that romantic love doesn't end. Instead,

* I'll discuss this more when I come on to talk about attachment theory.

like most psychologists, most of the people I spoke to in long-term relationships talked not of a never-ending romantic love or one that dipped in the middle, but of one that eventually turned into companionate love.

Although most people were candid about a number of controversial topics, some asked to go 'off the record' before conceding, 'Well … passion, romance, they do go over time – it's more about friendship in the end.' I wonder whether their reluctance to admit this reflects an expectation that romantic love should last forever, when in reality, it often turns into companionate love.*

I also wonder whether this transition from romantic to companionate love is more than just a shift: maybe it's a loss – a loss of the exhilaration that accompanies lust and romantic love, the sense of being alive and of being physically attractive and youthful. It may also involve the loss of dreams that one relationship could provide everything you'd ever hoped for. Perhaps the transition from romantic to companionate love requires both self-awareness and an understanding of love, as well as the space to mourn its unavoidable insufficiencies.

Love at first sight

'Then I saw her. I fancied her immediately and we spoke for half an hour at her window. She doesn't believe me, but I knew at the window that I would marry her.' Ricardo, who lived in Colombia, knocked on Paula's window because he thought he might be able to get her vote in an upcoming election.

'It wasn't love at first sight for her – she very quickly decided that I was a fool – but it was for me. I was 19.

* I'll talk more about this in Chapter 7.

'I asked her out, but she repeatedly said no. We lost touch, but I regularly thought about her … A couple of years later, a friend of mine called Jane posted a photo of Paula on social media.

'I immediately sent Jane a message saying, "If you actually know Paula, can you see if she remembers me? I am the guy who asked her out loads and she said no." Jane wrote back instantly and said, "Paula's sitting next to me – she wants to know if you are going to the party next week." And I thought, "Well, I'm definitely going now."

'I had a very physical reaction when I saw Paula again. It was like being back at the window. I thought, "Wow, she looks identical 12 years on." I was very drunk and talked at her for an hour. I sent her a message the next day saying, "I'm sorry for talking at you – shall we meet for a drink?" and she said, "Yes." Finally, she had agreed to a date.

'In the first few months with anyone, you think it's the best thing ever, when actually, it might fall apart because you're trying to change who you are for the other person. But now, years on, I think it's acceptable to feel smug. I am able to be exactly who I am with Paula and she accepts me.'

Ricardo wasn't the only person I spoke to who fell in love at first sight. Linda, in her 60s and happily married for 26 years, told me that her husband 'was in one department in the factory and I was in another, and every time I seen him, my heart used to go wild … That is no word of a lie. And I just thought, "He is mine," and we have been together ever since. We met in the February and we married in the September.

'He does everything for me, he cooks, he does the washing, he will do the ironing and then if he is at work then it's vice versa, I do the cooking and cleaning, but we work as a team all the time, we are just as a team. We give and take with one another, if we have an argument we just sort it out, we don't carry it on for weeks or days or hours, we say what we have got to say, forget it and move on.

'My mum and dad's marriage never lasted and my sister, she got divorced as well, so I feel quite proud actually. I have done my mum proud for 26 years of marriage.'

But for every story with a happy ending, I heard another where someone was convinced it was love, only to discover they'd got it very wrong. Like Dani, who ran a small business in the south of France.

'It wasn't love at first sight, it was love on the third date – but it was like being on drugs. I can remember the feeling so clearly of looking up at the right side of his face and thinking, "I will love you forever," and feeling chemicals coursing through me and going "this is it". Sad really because it wasn't true.

'I fell in "love" with someone who wasn't even the age he said he was, let alone the personality that he portrayed. Everything about him was a lie and there is a little bit of me now that thinks, "What does that say about me, that I fell in love with this mirage?" But it was very real. The physiological, chemical and psychological reaction was completely real, but I might as well have fallen in love with a pop star, that's how pointless it was. So, a sad tale really.'

Dani's description of the 'physiological, chemical and psychological reaction' that was akin to taking drugs reminded me of others' accounts, of rats opting for starvation to stimulate the brain's reward system and of Stendhal's description of crystallisation. But what struck me as dangerous was that Dani had no idea that her 'love' wasn't real.

I asked her whether, at the time, there was any part of her that doubted how she felt. Without pausing, she said:

'No. I couldn't imagine not ever loving him. I remember thinking that I should write a will so I could leave everything that I owned to him … In no time at all, I was just thinking, "This person is going to be the most important person in my life". Not that he would ask me for money, thank God, because I would have given it to him, all of it.

'Everything he said just kept on building an admiration and interest in him alongside a big physical attraction which created this mental explosion. I remember thinking, "I don't want to ever change anything about you" – even though he wasn't perfect (he was receding and had gammy eyes). I don't know what it means when you fall in love with someone who isn't real, but it really sucks. Don't do it.'

Thankfully, before giving away all her cash, Dani realised this wasn't love at first sight. After a few years of being single, she went on to marry someone she hadn't fallen for immediately.

Lust at first sight

I was really hoping to find some statistics about the percentage of people who fall in love at first sight only to realise they were wrong. But, as far as I am aware, no such study exists.

One of the few studies I did come across on the subject found that people who fell in love at first sight went on to have the same relationship quality as those who were friends first. But this may not actually tell us that much because the researchers only looked at those who were still together. Many of those who had fallen in love at first sight may have broken up before the study began.

As I mulled over the evidence, I came to wonder whether love at first sight was actually lust at first sight (albeit, with the potential to turn into love). I talked about this with Blackburn, who said: 'The right thing to say is that when it goes well and everything's fine, it turns out to have been love. That is, love requires a certain timescale and a certain history. And if the whirlwind doesn't lead to that timescale and history, it wasn't love, it was just a delusion or a fantasy of the moment. And you only find that out with hindsight.'

Adam, who I met in Antarctica and who had sold all his belongings to travel the world, had a similar view.* He argued that, while some people who fall in love at first sight turn out to be right, they may just as easily have been wrong: 'If they have chosen wisely ... they happen to have been right in the infatuation phase – but that doesn't mean that they weren't overconfident to begin with.'

In light of this, he advised postponing any big decisions until the 'drugs' wore off. 'I wouldn't want to make life decisions when on heroin. I'm sure it's wonderful, just as dopamine is, but it's not the best time to make life decisions when you're high on either.'

I have to say, I think I agree.

---

*You'll hear more from Adam in Chapter 5.

CHAPTER THREE

Heat-Seeking Missiles

Attachment theory* is one of the most researched areas of psychology and a complete game-changer when it comes to understanding relationships. The basic premise is that we're not all the same when it comes to intimacy and commitment. Instead, we each have a relatively consistent 'attachment style', which has a huge impact on our romantic relationships.

This, the theory claims, is largely down to our upbringing. But it can also be influenced in later life by our adult relationships, seeing a psychologist or suffering trauma (including the death of a loved one).

Everyone is somewhere on the 'anxious' scale and somewhere on the 'avoidant' scale. Secure people, who score low on both, are usually loving, communicative, feel comfortable with intimacy and tend to ask for help when they need it. They expect others to be loving and responsive, behave in ways that enhance (rather than damage) their relationships and don't worry too much about them.

People with an anxious attachment style,† on the other hand, are usually preoccupied with their relationships. They like to be close to their partner, worry about whether their partner cares about them and are often viewed by others as 'needy' or 'clingy'. They also can exhibit what psychologists call 'protest behaviour', which usually means

---

* Not to be confused with 'attachment parenting', which is very different (and is *not* one of the most researched areas in psychology).

† From now on, I will occasionally refer to 'anxious people' for the sake of brevity, but I mean people with an anxious attachment style rather than people who are anxious in the traditional sense of the word.

getting angry or frustrated and not wanting to be close to their partner (which they do to get their attention). Finally, they're very sensitive to threats because they're less able to regulate emotions.

People with an avoidant attachment style are also less able to regulate their emotions, but in a very different way – they simply don't connect to them. This means they don't (consciously) worry about relationships or rejection (they are more likely to be doing the rejecting). They try to be as independent as possible and, when in a relationship, try to maintain distance. They can be arrogant, unforgiving and picky. They can also idealise people they're not with and are often attracted to relationships that are unlikely to last.

Those with an unresolved attachment style don't have a coherent attachment strategy – instead they can oscillate between avoidant or anxious. They've often been maltreated as children or suffered trauma as an adult – but haven't processed their experiences. As a result of this unresolved trauma, they can find themselves surrounded by triggers and are often disconnected from their bodies and emotions. They are usually classified with a secondary attachment style depending on which they adopt the most.

To give you a sense of the numbers, when researchers analysed more than 200 studies involving over 10,500 adults, 58 per cent were found to be secure, 23 per cent avoidant and 19 per cent anxious.* And these numbers were largely independent of language or country.† That means that a total of 42 per cent of people across cultures are

---

* The different attachment styles are called a variety of things, but in the interests of clarity I'm going to stick to these names. I should add that in this study, 18 per cent were actually classified as unresolved, but given a secondary classification.

† Incidentally, other studies have found that attachment isn't related to intelligence or memory – so no one is too clever to be insecurely attached.

insecure.* So even if you yourself are secure, at some point in your life you will probably find yourself dating someone who is not. Yet attachment is not something that is on many people's radar.

Friends aren't likely to point out that the reason you worry your partner's cheating on you might be that you are sensitive to threats thanks to your highly anxious attachment style. Or that you've broken up with yet another person because you're subconsciously looking for reasons to maintain your independence and protect yourself against loss. Or that you might be in a destructive relationship because that's what feels familiar to you based on your upbringing. People don't tend to have the knowledge or understanding of attachment theory to have these conversations. And even if they did, they may not want to hear it – as well as being subconscious, the attachment system can be compelling and potent.

People don't generally ask about the attachment style of the person they've just started dating. Alongside location or personal interests, dating websites don't have a box for 'securely attached' or 'insecure but working on it'. Yet it is likely to play a much bigger role in your romantic relationships than, say, hobbies or height.

I'm going to go into more detail about the different types of attachment style and how they impact romantic relationships, but before I do, I want to explore the 'attachment system' and why psychologists think we have one.

From the cradle to the grave

The founder of attachment theory, John Bowlby, believed that our attachment behaviours evolved to help us survive,

---

* 'Insecure' is an umbrella term to describe avoidant, anxious and unresolved attachment.

particularly when we are young. It's easy to imagine how useless a six-month-old baby would be at protecting itself from a vicious tiger, so it makes evolutionary sense to form 'attachments' to people who will (hopefully) protect us in times of danger and ultimately help us survive.

The driver of this, Bowlby argued, is our innate 'attachment system', which he likened to a heat-seeking missile. Our attachment system establishes its target, then drives us to do whatever's necessary to get close to that person, while looking out for any changes in their behaviour that might mean we have to adjust what we're doing and how we go about trying to get close.

As our parent* responds (or doesn't) to our heat-seeking missile's attempts to get close, we begin to build an internal model that becomes our template for adult romantic relationships. We learn that others will – or won't – be there when we need them. In some cases, we might learn that others will get angry with us – or even hurt us – if we ask for their support.

We then use this model to make subconscious predictions about how both we and the significant people in our lives will behave in a relationship. It changes the way we see ourselves and others, the way we focus our attention and the way we remember things. In other words, our attachment system is ridiculously influential – it changes the way we see, experience, remember and think about life. It becomes our way of being in the world – our 'truth'.

---

* I refer to 'parent' as shorthand for whoever's responsible for a child. But kids can form attachment bonds with grandparents, guardians, siblings, day-care providers or occasionally other children – as well as, or instead of, their parent(s). Kids usually develop two or three attachment bonds in their first year and there's always a hierarchy. They might be able to cope with a major separation from a less important attachment figure, but the same sort of separation from their main attachment figure would devastate them.

Bowlby believed our internal models could be updated by developing an awareness of them or experiencing new relationships, but at the same time he thought they were robust and resistant to change, famously saying they stayed with us from the cradle to the grave. This is in part because they are not conscious – but also because they're based on incredibly powerful self-protective defence measures that don't give up easily. More recent research suggests that there are things that insecure people can do to help themselves become secure, but I'll explore that towards the end of the chapter.

Bowlby's theory was soon backed up by countless studies. Psychologist Mary Main found that attachment styles were consistent from infancy up to 19-years-old more than 80 per cent of the time – but only if people who had suffered trauma (e.g. the death of a parent) were ignored, because trauma can have a massive effect on attachment. A more recent review of 27 long-term studies* found a weaker consistency in attachment styles over time – but they were still fairly stable.

Attachment theory isn't without its critics. Some argue that it underestimates the importance of peers in a child's development, or that it's a ploy to justify policing mothers by making them solely responsible for their children. Others claim that it's heavily based on research conducted in Western cultures – but this could, and should, be said of a huge amount of psychological research. In any case, studies have often found the same distribution of different attachment styles across cultures.

I'm pretty convinced by attachment theory and the enormous body of evidence that supports it, but these

---

* I say 'long-term', but the correct term is really 'longitudinal'. A longitudinal study usually involves repeated observations of the same people over time. They don't actually have to be that long-term; the key aspect is the repetition over time, even if that's just a couple of months.

critiques are valuable as a reminder that our attachments are only one of a number of influences on our lives and that they are affected by external factors – for example, parents are likely to do a better job of cultivating a secure attachment if they themselves are supported.

If attachment theory is right, humans have not evolved to operate independently. Instead, we're programmed to develop attachments in order to survive – as an infant, attaching to someone is literally a matter of life or death. As a result, the attachment system is incredibly powerful, capable of driving our thinking, memory and behaviours, but at the same time it lies beyond the reach of reason and consciousness. That's a dangerous combination.

What's more, much of our attachment learning happens before we develop language skills or conscious memory. So, we probably won't be able to explain why we act the way we do and, in the case of insecure attachment, will simply find ourselves driven by an invisible – and hugely powerful – force, one that can make it more difficult to have intimate and committed relationships.

A shield of armour – secure attachment

As I've mentioned, there are four different types of attachment style: secure, anxious, avoidant and unresolved (anxious, avoidant and unresolved being 'insecure').*

People usually develop a secure attachment style when at least one person in their life provides a 'safe haven' in times of crisis and a 'secure base' from which to explore the world. This generally happens when a parent is sensitive to the way their child communicates and meets their needs quickly and consistently. It doesn't mean getting it right all

---

* Although it's worth keeping in mind that individuals and their relationships are unique and this sort of clean and neat classification can never fully encompass their reality.

the time – the key is to be 'good enough' at being warm, responsive and reliably available.

Secure people learn that if they ask for support, they get it, and come to believe that they can influence the world and achieve things through their own effort. They also learn that they can count on others. They are more likely to explore and be confident and independent knowing that they have a safe haven when they need it.

A number of people I spoke to recognised the importance of this sense of safety – even though they hadn't heard of attachment theory. Lia, who was in her 50s and from Brazil, said: 'When my parents were alive, I felt safer because, no matter what, I could always, always go back to them ... Thanks to the love of my family, I have always had a feeling of belonging. And it doesn't go away. I could live anywhere in the world, but the feeling that I am not misplaced in life is there.'

Harry, a student in her early 20s from Denmark, said: 'Love is about fun and safety. It's about always knowing that people are there if you need them. When we're growing up, we're designed to look for safety and love. I don't think a lot of people understand the importance of safety.'

And Anthony, a 30-year-old from London, said: 'Good parenting is about building confidence in a kid so that when they are an adult, they go out into the world with a shield of armour: they are safe. They believe they are amazing because that is what their family has instilled in them. If parents don't give a child that shield of armour, the child will be insecure.'

The benefits of secure attachment (and the disadvantages of insecure attachment) are enormous. In fact, some psychologists have mused whether there is anything that attachment is *not* related to.

Secure children are less fearful and angry, more joyful, have higher self-esteem and more faith that they'll be able to cope with suffering, are less threatened by stressful events, more persistent and are less likely to develop mental health problems. They also have a positive but realistic sense of self and are capable of admitting that they are imperfect, whereas insecure children have a more negative self-image (anxious) or resist admitting flaws (avoidant).

As you can probably imagine, having a secure attachment style is really helpful when it comes to romantic love. As adults, secure people are comfortable with intimacy and commitment, ask for help when needed, expect others to be loving and responsive, and don't worry needlessly about losing their partner. Whether they're facing a missile attack or chronic pain, relationship conflict or a break-up, secure people look for support informally through parents and friends, and formally through teachers and counsellors. This programming runs very deep and means that, without realising, secure people pay less attention to threats and more attention to loving cues – which obviously makes for better relationships.

They are also more likely to be in a relationship than insecure people and tend to be happier once in one. This might be because their relationships are generally more stable, satisfying, intimate and committed, and they have more positive (but not unrealistic) expectations of their partner. Oh, and – in case that wasn't enough – they generally act in ways that boost (rather than harm) their relationships.

The benefits of secure attachment are staggering – and maybe even enviable for those who are insecure. Alexandru from Romania* said: 'I would love to be securely attached and know people who are … I am sure there is some club of these people, they probably don't know who they are even, which is sad. Securely attached people basically attract securely attached people, the whole thing just flows very smoothly for them and it's all fine, and it's fine that it's fine and they don't know it … I just want to be in the securely attached world.'

* You'll hear more from Alexandru later in the chapter.

But it's not *all* safety and joy for secure people. One possible downside for them is that, as a romantic partner, they may lack drama at the outset because they communicate, are open and responsive, and are comfortable with intimacy. This means that they are much less likely to trigger their partner's attachment system in the way that anxious, avoidant and unresolved people can – and some people may interpret this as boring, settling or lacking in 'chemistry' or 'spark'.

Insecure attachment

Given its advantages, having a secure attachment style is obviously the ideal – but sadly not all parents are available, responsive or consistent. And, as children, we're biologically programmed to develop attachments, even when people aren't actually very good at making us feel secure. In extreme cases, we form attachments to people who neglect or abuse us. This can lead to 'insecure' attachment. As psychologists Chris Fraley and Phillip Shaver explain, if a parent is cold, rejecting, unpredictable, insensitive, inconsistent or even frightening, children learn that they can't rely on other people for support. Of course, parents don't operate in a vacuum – they're hugely influenced by their circumstances and some simply aren't in the best position to build secure attachments. Those who live in chronic poverty, are abused by their partner (or anyone else) or have mental health difficulties may not be able to respond to their children with the sensitivity and consistency necessary to nurture security.

Parenting isn't the only path to insecurity. A child can develop an insecure attachment after the death of their parent, abandonment or their parents' divorce. Even threatening to leave can disrupt their attachment style, as can the lead-up to a divorce – if children see their parents withdrawing from each other, they might worry that something will happen to one or both of them. Being exposed to parental abuse has a similar effect, after all, this involves a child being exposed to the threat of one or both of their parents on a regular basis. And it happens with worrying frequency – in a UK study of more than 6,000 people, 24 per cent of 18 to 24-year-olds admitted to being exposed to parental violence growing up.*

Rather than learning to be independent, insecure children internalise a paradox: they must rely on others, but at the same time, these others may not be there for them. They learn that if they ask for support, they may or may not get it and when they try to achieve something they often fail.

But there's hope. Studies suggest that insecure people can work to develop a secure attachment style. I'll come to this a bit later.

I just love my own space – avoidant attachment

'I don't think I'm very good at romantic love.' Diane was in her 50s with chestnut hair and a hearty laugh. We met while walking in the Scottish Highlands, and spoke surrounded by gnarly tree barks, mint lichen and bright daffodils.

'I have good relationships, but they don't really last – and it's usually me who ends them. Maybe they last as long as they need to, which is five years on average.

'I can think, "I am such an adventurer, I've had all these adventures … " And then on a bad day, I think, "Oh God, here we go again … What's wrong with me? I can't sustain a romantic relationship … " It depends on my mood, on how I wake up that morning.

---

* I use the word 'exposed' because children can be aware of parental violence without actually seeing it take place (although plenty do see it). Children might hear it, for example, or see bruises, cuts or broken furniture after the incident(s).

'I just love my own space. I'm in a relationship just now – but I don't think I could give up my own space. I like to be able to wake up in the morning and think, "Here I am in my own space, I can have no music – I can have music. It's my choice. I don't have to negotiate anything." And when I feel like I want to compromise, it's lovely to see him. I like it when you're on the same page and think, "Okay, I want to negotiate today – come round."'

As Diane spoke, I wondered whether she was avoidant – she tended to break up with people and she mentioned the word 'space' three times in three sentences. But I didn't want to jump to any conclusions, particularly as researchers only decide on an individual's attachment style after an intimate interview or detailed questionnaires, so my assessment was hardly scientific. I asked, as I often do in interviews, whether she knew anything about attachment theory – and she did.

'I've studied it. It was part of my professional training as I work in child protection – and I'd say I have a bit of avoidance in me.

'I think it's down to my mother. If I wanted nurturing, I would go to my dad because he was softer – but she was the main carer. I don't want to say she was cold, because she was great, but if I fell over, it wasn't comfort I got. It was, "What the bloody hell are you doing down there? Get up … You'd better not have made a hole in your tights." There was a harshness to it. Whereas my father would say, "Oh no! Are you Okay?"'

Avoidant attachment

As children, people with an avoidant attachment style learn that being upset leads to rejection (or worse), in Diane's case, in situations where she would ideally have had comfort, her mother told her off. So, in order to prevent both, avoidant children try to stop themselves from activating the attachment system altogether. This stops them from looking for love, support or comfort in the first place. The problem is that this comes at a high price – in order to disconnect from their attachment needs, they disconnect from their emotions and, often, their bodies.

As adults, highly avoidant people don't value relationships and, as a result, they're less likely to prioritise or put any work into them. Instead, they idealise independence. They try their best to avoid relying on others by being obsessively self-sufficient, rejecting help, avoiding emotional involvement and denying their needs for love or comfort. They don't connect with, let alone verbalise, feelings or thoughts that might encourage them to look for support from others. This usually means they do their best to avoid negative emotions, weakness and vulnerability. Unsurprisingly, this severely restricts their self-awareness (which can leave them unaware of their impact on others) and compromises their ability to empathise.

But what they try to avoid in themselves, they can bring out in others. For example, by trying not to need their partner in any way, as well as distancing themselves from them, they're likely to make their partner feel vulnerable – which is exactly what they subconsciously want to hide from in themselves. They also unknowingly overestimate their own value and can appear arrogant,* and they find the negatives in others so that they can reject them (often projecting on to others what they don't want to see in themselves). In other words, they can be unforgiving or picky.

Focusing on imperfections is something that Tristan (who I met in Switzerland) and Alexandru both admitted to. Tristan explained: 'I absolutely, without a doubt, live in

* Because personal weakness threatens their only source of protection – themselves.

a fortress … and tend to be very unforgiving. I turn things into red-line issues when they really shouldn't be and I say to myself, "I don't want to be with someone who would say that or who would do that." It's a failure to adjust to the complexities of someone else's character.'

And Alexandru admitted: 'When I meet new people, often I will think they are not good enough, like, "She is not totally what I want and so I can't date her any more." In my head, this is 100 per cent what I believe. It's not a lie that I am telling myself that I can confess to later. I really believe that the person who is in front of me has some imperfection that completely disqualifies her from any intimate relationship at all, which is crazy.' He added that, as an avoidant, 'You're not available anyway, so when you are saying there is no spark with someone, the way that you relate to other people is broken. You could meet Beyoncé and still find faults.'

Another common strategy is to idealise others. They usually do this with people they are not going out with or can't go out with, including people they have already broken up with, who are married* or live abroad. Alexandru explained: 'All my friends will tell you this, every girlfriend that I have ended it with, I've had regret and thought, "Should I get back together with her?" and "She was so perfect and there will never be anyone like her." With all of them, I tried to get back together in some stupid way and then obviously when we are back it's like, "Oh my, I can't be here."'

Avoidant people can idealise partners they've broken up with because their defence strategies lift and they can see the positives more clearly – and then exaggerate them. But

---

* Remember Claire who you met in Chapter 1? When she dated married men, part of her wanted them to leave their wives, but at the same time, she recognised that she probably didn't want the commitment either.

if and when highly avoidant people get back together with their ex, their avoidant defence mechanisms usually kick in again and they often start focusing on the negatives as they did before.

That said, being avoidant doesn't mean *never* being in a relationship – but the relationships they do have usually lack intimacy and can feel very suffocating for them, often to the point where they have to leave. This is particularly true when they date very anxious people (which they frequently do), who are preoccupied with being close (when what an avoidant person wants is space).

Alexandru explained: 'Avoidance means that you are very prone to feeling trapped very quickly and … relationships are a minefield for this because whenever anyone asks you to do something, whenever someone comes into your life in any way, that's the first thing that you feel.'

He paused to think about his last relationship, before admitting to 'telling myself, "She's trapping me" and having these fantasies about stuff that I could do, like, "Maybe I should go to business school to escape" … by the end I just had to get rid of her, I felt claustrophobic and angry and resentful and I just had to walk away from that relationship. We didn't talk. I lacked the words to really talk to her about it and it was just "goodbye".'

He wasn't the only one who had felt claustrophobic before needing to escape. Juliet, from Australia, told me, 'When I am dating someone, I tend to wake up one morning and change my mind and I can't explain why.' And Terri, from Arizona* in the US, explained: 'I definitely see a lot of my mum in me. She tends to pull away and I notice that I do that sometimes too … I get really restless

*You met Terri in Chapter 2 – you might remember she likened lust to a firework, all BANG and BOOM …

and lash out, and I don't know why. Everything he does annoys me and I can't stand being around him, which is unusual because he is my favourite person to be around. So, when it comes to times like that ... I will try and go off and do something on my own or turn my phone off for a little while. It annoys him sometimes, but I'm like, "I just need to get away from you so that I can love you."'

Other subconscious avoidant strategies can include: thinking they're not ready to commit (but staying with their partner for years); flirting with others (which brings insecurity into the relationship or keeps other options open); forming relationships that are unlikely to last (e.g. with people who are married, live abroad or are emigrating); keeping secrets; avoiding physical closeness; checking out mentally when their partner is speaking to them; and not saying 'I love you' (despite having or implying that they have feelings for their partner).

Most people I spoke to thought avoidance was a particularly male way to behave, whereas women were far more likely to be anxious – but the evidence doesn't back this up. Adult and infant studies internationally have found no gender difference – there are just as many anxious men as there are women – and the same is true of avoidant attachment.

The real story

Avoidant people might minimise the importance of love and idealise independence, but that's not what they actually believe – study after study has found that they're just very good at repressing their needs when it comes to love. From adults in divorce courts to children who have been left with a stranger in an unfamiliar place – people who are more avoidant can appear calm and collected, but their bodies tell a different story. When being interviewed to assess their attachment style, avoidant people don't usually talk negatively about their parents – in fact, they are often very positive. But 'skin conductance' tests* suggest they're having an emotional reaction.

In one set of studies, people were asked to report words shown to them on a monitor. Those with an avoidant attachment style were slower than others to recognise words like 'separation', 'fight' and 'loss' – until they were distracted by a puzzle. The distraction reduced their ability to repress and they reported these words as quickly as non-avoidant people.

Repression might work in a lot of day-to-day situations, but it only works up to a point. When faced with a very stressful life event, such as military trauma or the birth of a severely disabled child, the defences of avoidant people are quick to break down and they behave like anxious people.†

Avoidant people do crave connection – specifically, a safe and secure relationship – but instead they tend to live on a restrictive and punishing diet of self-sufficiency, idealising people they're not with and finding negatives in the people they do date. A highly avoidant person's response to an intimate relationship is like that of a starving man at a banquet who tells himself that the food simply isn't good enough, and so refuses to eat.‡ The problem is that, most of the time, avoidant people have no idea what's really driving their behaviour, or its impact on others. They just believe that the food really isn't edible.

---

* 'Skin conductance' is when skin momentarily becomes better at conducting electricity as someone is 'physiologically aroused'. The idea is that if they're physiologically aroused, it's likely to be thanks to emotion.

† Which is tough because they become the polar opposite (dependent) of what they idealise and strive to be (independent).

‡ I owe this analogy to psychologist David Wallin.

I'm very needy – anxious attachment

'Anxious attachment as a child felt like I had something heavy on my chest when other people weren't around me and so I constantly used to hug my mom or my sister. Or when I was sleeping, I would wake up and feel anxious so I'd crawl into bed with my sister and touch her arm so I could feel body contact. It just gave me such a sense of security. I always did that with my mom, constantly climbing all over her and sitting on her lap or just sitting anywhere near her. I think people thought that was endearing as a child because it gave them attention.'

Sarah was 23, American and disarmingly honest. We sat on an old wooden bench as she told me about her upbringing. Her parents divorced when she was three-years-old and her mother, an alcoholic, remarried a year later. But that ended too. Sarah remembered her mother waking her up at 2.00 a.m., telling her they had to leave immediately and throwing all her clothes into bin bags. She remarried 18 months later and by the time Sarah was 10-years-old they had moved 10 times.

'As I got older, I tended to pick men that weren't affectionate physically, or even sexually, not the type to instigate holding hands or hugging. I was so afraid of looking needy that I would repress the urge to be really affectionate. I would pretend that I didn't want their affection and try to seem like I was not bothered if they hugged me or not, or gave me a cuddle in bed, but deep down I always would want it. It made me feel like I wasn't being my true self. And any time you feel like you are not being your true self, it gives you anxiety. So physical closeness is important for me as someone with an anxious attachment style.

'Also, in relationships, if I feel there is any chance that the power dynamics are off, it throws me through a whirlwind of anxiety. If I feel like they are playing games with me, waiting six hours to text me back purposely, I will play games with them too. I will feel like, "Okay, well, if that is the way that you are going to treat me, I am just going to throw it back at you 10 times as hard."

'I have these thoughts in my head, but I never really articulate them. And boyfriends have no idea what's going on, they don't even know that I am playing these sorts of games with them. Then when I see them in person and I realise that they weren't necessarily playing games and might have been busy, I am completely at ease. It's when we're away from each other that my anxiety rises. When we are in the physical company of each other I feel really secure most of the time – if I am happy in the relationship and there aren't any issues, and if my partner is the type to be affectionate.'

Sarah's experience resonated with everything I'd been learning about attachment theory. As children, anxious people like Sarah are driven to seek protection from the people they're attached to. They learn that their parents are more likely to respond if they cling to them, make contact, call or cry – and in most cases, this works at least some of the time, so they keep doing it. As it's consolidated over years and years, this behaviour becomes very hard to change.

As adults, highly anxious people talk about their fears, needs and doubts, continually worry about their partner's availability and are very sensitive to threats. They can get very upset when they're apart, so frequently try to be close to their partner as often as possible. This is very different to the way that avoidant individuals experience relationships: anxious people crave closeness, avoidant people crave space.

Sophie, a model I sat next to on a flight to Spain, gave a great insight into what it felt like to be apart from her boyfriend. She spoke as if she was in a rush to get somewhere (when she was actually stuck talking to me on a plane).

'When I'm not with him, all my insecurities start playing out, I start thinking why it shouldn't be working, what he could be doing … I'm very needy. I just want someone to call me all the time … I can be a feeler, but when that person's not around, I'm just thinking about the feeling, think, think, think, think, think. I'm overanalysing what they've done – why have they looked at me like that? etc. I just bought some self-help books because I overthink everything – it's like I'm in a rush.'

One of the reasons anxious people get so stressed when they're apart from their partner is that they haven't learnt to soothe themselves or regulate their thoughts or emotions. In one study, researchers asked people to suppress thoughts about a romantic partner leaving them for someone else, while continuously writing down any thoughts and feelings they had. They then assessed people's ability to suppress these thoughts by examining their physiological arousal and the number of times these thoughts appeared in writing. Anxious people were less able to suppress thoughts about abandonment compared with those with an avoidant attachment style.

In another study, people's brains were scanned as they were asked to think about, then stop thinking about, different relationship scenarios. Researchers found that, in anxious people, the areas of the brain related to emotion were more activated and the areas related to regulating emotions were less so. Not only were they more prone to react to thoughts about loss, but they were less likely to regulate their emotions once they had reacted.*

---

* I have to be careful when making generalisations about the way brains process emotions, as recent research suggests that there is no 'fingerprint' for each emotion in the brain, as psychologists once thought. And while brain scans provide a useful insight into the brain, they have their limits, and don't necessarily capture the full and astonishing complexity of brain function.

It is not only thoughts about relationships that can lead anxious people to think about loss; negative thoughts that have absolutely nothing to do with the relationship can trigger worries about rejection or abandonment. Put simply, highly anxious people live in a world full of triggers, and this means they spend a lot of time and energy worrying about relationships.

They can't be calmed until they feel secure and they know that their partner is there for them. They can therefore spend a lot of time with their attachment systems chronically activated – particularly if their partner is very avoidant. The good news is that their attachment system can be calmed down relatively quickly and easily – sometimes it might take no more than a quick message, phone call or hug from their partner if received early on. But if left to escalate, it will be much more difficult to calm them.

Anxious people can also exhibit what psychologists call 'protest behaviour'. This usually means getting angry or frustrated and not wanting to be close to their partner, when their actual aim is to get their attention so they can be close to them (even though it can have the opposite effect).

In adults, protest behaviours might include withdrawing (e.g. by becoming silent, physically turning away or ignoring their partner); keeping score (e.g. deciding not to make contact with their partner again until they call); being hostile; threatening to leave; pretending to be busy or ignoring phone calls; calling or messaging repeatedly, trying to 'bump into' their partner; and trying to make their partner jealous (e.g. by meeting up with an ex or telling their partner how other people are interested in them).

Sarah explained what protest behaviours were like for her: 'I oscillate between saying, "If they are not going to put in the time in this relationship I don't even care about them and I can get anybody that I want anyway." If they are not bothered then I am not bothered and I put up this gigantic barrier in my own head. Then on the other side it will be, "Maybe that last message I sent to them was a bit mean, maybe I need to soften my approach some more." So maybe I start looking at myself and the way that I am responding to certain situations and start being really hard on myself. But it's usually a mixture of the two: me being ostensibly hyper-confident about myself – "I can get anybody that I want so why even bother with this person if they are not responding to my needs in this way" – and then switching to the other side, "What did I do to deserve this?"

'As I am learning more about attachment theory, it is easier and easier to recognise what I am doing and try to have more rational thoughts about it. Ultimately, I think the thing that gives me most anxiety is my own head and the games that I play with myself.'

Fear without escape – unresolved attachment

The last of the three insecure attachment styles is unresolved. Unresolved people don't have a coherent attachment strategy. Instead they can oscillate between behaving like anxious and avoidant people – and are generally given a secondary classification based on which attachment style they adopt the most. Like anxious and avoidant attachment, it often makes it difficult to have intimate, committed and happy romantic relationships.

Unresolved attachment is as common as avoidant attachment and even more frequent than anxious. One review of studies looking at more than 2,000 children found that 15 per cent were unresolved (compared with 9 per cent anxious, 15 per cent avoidant and 62 per cent secure).* I should point out that a lot of this research was

---

* Psychologists call this 'disorganised' attachment when talking about children and 'unresolved' when talking about adults. But in the interests of simplicity, I am going to use 'unresolved' to refer to both.

conducted in middle-class environments – other studies suggest that children from lower socio-economic groups are even more likely to be unresolved.

The causes of unresolved attachment are complex; there's rarely one simple explanation and it can stem from a number of experiences. A common cause is parental child abuse, as this puts children in a paralysing contradiction – they naturally want to turn to their parent for safety, they're dependent on them for survival, but at the same time, they want to run away. They're frightened, but with no solution or escape.

'My stepdad used to threaten me with burning my face with a hot iron while I slept if I told my mum he was beating me up,' explained Oscar, a photographer from South Africa.'When kids were beating me up at school, his advice to me was, "Wait until their back is turned and hit them over the head with a pole." I knew that he was a coward and I was scared that when I wasn't looking he would beat me up. He was a loving guy when he wasn't drinking, super cool like Marlon Brando, drove a motorbike – a rock and roll dude. But one time he thought that I slammed the doors, so he beat me up – it was the wind. When I spoke back to him he threw my radio, the only thing I loved, and broke it against the wall next to my face.'

Child abuse is not the only cause. Children can develop an unresolved attachment style when their parents act as though they're scared of their children, look to them for comfort or go into a trance-like state.

This frightening, frightened or disconnected behaviour often happens in parents who've gone through trauma but haven't processed it. They may have been abused as children (physically, emotionally or sexually), or been traumatised as an adult (e.g. suffering partner abuse, rape or losing a loved one). Psychologists call this 'unresolved trauma'. And the 'unresolved' part is key – the fact that it's unprocessed or buried means that it can bubble up and take hold of them, leaving them petrified, in a trance or in a rage. This sort of behaviour, which can be triggered by their children's actions, among other things, leaves their children vulnerable to developing an unresolved attachment style. Or at least, that's the theory.

As well as being frightening, frightened or disconnected, a parent with unresolved trauma may be too emotionally unstable and inconsistent to offer much comfort and protection to their child – which is what children need to be secure.

Trauma and abuse

I wanted to cite a few statistics here just to give you an idea of how common abuse is. In a national study in South Africa in 2001, 29 per cent of women and 21 per cent of men had been physically abused in their most recent relationship; and in a 2006 UK study, 35 per cent of men and 40 per cent of women had experienced partner violence in a same-sex relationship.

As for child abuse, in a 2009 UK study of more than 6,000 people, 6 per cent of under 11-year-olds, 19 per cent of 11 to 17-year-olds and 25 per cent of 18 to 24-year-olds had been severely maltreated during their childhood. Maltreatment was defined as including physical, emotional and sexual abuse, as well as neglect.

Child abuse came up a lot in my interviews – I spoke to people who had suffered physical, emotional or sexual abuse at the hands of parents, neighbours and even kidnappers. We'll come back to some of them later in the book, but for now, I want to talk about Neshbet, from Turkey. He was in his 40s, about 6ft tall and had a skinny moustache. When he went to prison, he was quite judgemental of inmates, but quickly came to understand that many of them are 'people who have been abused from a very, very young age, having sexual relationships with close family members. Boys and girls, I've met both. They've had very bad lives.

'I talked to them. I said, "Why are you like that to other people?" And they said, "Why should I be different?" Their mother didn't love them, they don't know who their father is, they were abused sexually in care homes, and some of them moved care homes 25, 30, 40 times. They have never been touched with love. When they were suffering, nobody felt for them. You have to receive love to be able to give love. All they know is hate.'

The phrase, 'All they know is hate,' has never left me. And the more we understand the brain, the more we realise that Neshbet is right. When brain circuits fire repeatedly, they can become a default setting. That means that if, instead of love, comfort and protection, a child feels frightened and unwanted, their brain will specialise in managing feelings of fear and abandonment, rather than co-operating with other people and exploring the world.

One of the difficulties with trauma is that it can be more manageable to deny rather than face it – and this is something that maltreated children learn to do early on. A child has no choice over where they live and who they live with, so they have to organise themselves to survive within the families they find themselves.

Erasing awareness and developing denial are often essential to survival, but unresolved trauma can have disastrous consequences – both for the people who have suffered themselves and for any children they may go on to have (who may assume that they are somehow responsible for their parents' behaviour).

Unresolved behaviours

Unresolved children behave in ways that seem either contradictory or uneasy. They might, for example, have a fearful or dazed/trance-like facial expression, or walk towards their parent with open arms while looking away. And their behaviour tends to be specific to a particular parent

– in one study, only three of the 34 unresolved children behaved in unresolved ways with more than one person.*

Adults with an unresolved attachment style usually seem disorientated and disordered when talking about trauma they have suffered. They might, for example, fall silent mid-sentence and finish 20 seconds later as if no time has passed, or simply fail to finish what they were saying. Or they might have brief lapses in reasoning – so they may believe someone is both dead and alive, or speak in the present tense about someone who died 20 years ago. Psychologists explain this in the same way that they explain the frightened, frightening and disconnected behaviour – namely, that the buried traumatic memories come to the fore and take over the unresolved person's thinking and actions.

Although burying or denying trauma is a common and often necessary strategy, particularly in children, it leaves unresolved people disconnected from their bodies, cutting them off from vital messages about how they actually feel. It also leaves them with a very poor awareness of their own thoughts, as well as those of others. In the same way it does for highly avoidant people, this can make unresolved people feel remote, distant and unavailable, which is not helpful for romantic relationships or intimacy.

They can also be prone to explosive behaviour or irrational thinking when highly stressed and can find themselves living in a world of triggers. This leaves them repeatedly overwhelmed by memories of the trauma they've tried to bury – but which actually defines their way of being. And,

* I should briefly point out that psychologists measure children's attachment differently to adults'. A child is assessed by looking at one particular relationship – trained researchers observe them in a strange environment as their parent leaves and returns. Adults, on the other hand, are assessed more globally – either by very detailed interviews or questionnaires. So, although I talk about 'attachment styles', the accurate way to talk about it in children is to see them as having attachment behaviours specific to a particular relationship.

in the same way as it does for highly anxious people, this completely transforms their romantic relationships as it can be very difficult for them to feel safe with someone else.

Anxious–avoidant relationships

Your attachment style is largely down to your upbringing – but it can also be influenced by romantic relationships (as well as seeing a psychologist or suffering trauma).

One of the reasons they can play such a big role is that, over time, your partner's attachment style can impact yours. That can mean a shift from insecure to secure attachment when dating someone secure, but it can also work the other way if you're secure and date someone insecure.

Going out with someone with a very avoidant attachment style, who sees themselves as superior, feels claustrophobic in a relationship and doesn't want to spend time with you, may well make you feel less secure – even if their behaviour is not in any way personal or conscious. Meanwhile, going out with someone who becomes tense when you don't answer the phone, hates to be apart and can get angry when you are, may well make you crave space.

Avoidance is likely to be even more difficult to cope with if you're very anxious, and anxiety is likely to be even more difficult if you're very avoidant. Yet anxious–avoidant pairings are relatively common, perhaps because they each conform to what the other expects in a relationship – avoidant people expect to be less involved, whereas anxious people expect the opposite.* Anxious–avoidant pairings can also be long-lasting, but are rarely very happy. The

---

* I also wonder whether there's an element of expecting to have what your parents had (assuming you had two parents in your life) – and if you are insecure, that may well have been an anxious–avoidant relationship. But that's just me wondering and isn't based on any evidence …

paradox is that avoidant people crave space and independence, whereas anxious people crave closeness – and the two are usually incompatible.

One of the best insights I had into an anxious–avoidant relationship was from Alexandru. We sat next to a tall window, drinking herbal tea and watching the sky oscillate between overcast and sunny, as he told me about his break-up.

'It was a pretty serious relationship. We lived together at the time and it went so badly wrong it resulted in me behaving in ways that I didn't really understand or recognise in myself after we broke up … well, it wasn't really even a break-up, it was really a breakdown.

'I felt very policed. She ended up being very controlling with my time; she wanted me to be at home all the time. On one occasion she called a friend of mine that I was having drinks with asking where I was and it was that sort of behaviour that really drove me away. At the time, I thought, "Oh she's being so controlling and this is all on her," but I didn't ask myself what behaviours of mine were actually bringing that about.

'Being avoidant means that you are very prone to feeling trapped very quickly … I think that she definitely triggered me every day. It was all the time … I would have really inappropriate outbursts at her and that's a good way of checking your behaviour. If the way you are reacting is inappropriate to what has transpired in the moment, then you know that is something you should look into.

'I experienced a lot of anger and frustration that she wasn't automatically on the same page as me, which is totally stupid, in retrospect, if you don't communicate it. I just thought that a lot of things should be obvious, like, "Why are you insisting that I come home every night at seven o'clock?" I never took any interest in what led her to want that or asked her or engaged with her. I just stayed on my own side of the court and got angry. I'd start off being resentful, then as it happened more and more I felt full-blown anger. You can't feel too many emotions at once so anger buried a lot of affection and love that I had for her, and a lot of other stuff.

'It was a very hazy time and by the end I just had to get rid of her. I felt claustrophobic and angry and resentful, and I just had to walk away from that relationship and we didn't talk. I lacked the words to really talk to her about it and it was just "goodbye". I was too alienated by my own behaviour, my thoughts, my emotions during the time, it was just a side of me that I had never really seen and it all felt very compulsive towards the end of the relationship. The breakdown of that situation led me to try and explore the motivations of my behaviour.

'Frankly, my relationship wasn't that unique or special, it was pretty classic. I am avoidant and my partner was anxious. That's a very classic pairing.'

Alexandru went on to describe how his relationship unfolded for him. First of all, he thought she was going to satisfy the needs he thought he had, before realising that she didn't do that (because nobody could), and feeling betrayed and hurt that she was not the person he thought he was dating. Then he got angry at her, before it reached a point where he realised they had to break up or stay together but fight all the time.

'Or you stay together and avoid each other all the time and have separate lives ... But another option is actually to work through your issues.

'It seemed to me like both of us were going on our compulsive behaviours and were really just firing off each other in this automatic way, which left us in a lot of situations that we didn't know how to get out of. I don't think she was aware of what was going on and to my knowledge she has jumped into another relationship – which is a very anxious thing to do – so I doubt she has taken this opportunity to reflect on it. But I don't want to sound patronising here. We all just need to lead our lives, so I don't want to sound critical.

'At the time, I didn't have this vocabulary, and people are usually articulate about things that they have been educated about and are self-aware in those areas. That's what I found surprising about that particular relationship and that break-up, because I consider myself reasonably self-aware in other domains, in my work and my academic life. To then be confronted with the situation where I didn't know what to say, I couldn't really justify my behaviour and I couldn't analyse hers. It was just weird to be in a place where I completely lacked the vocabulary or the thought process. My neurons were not firing, I was in this compulsive state, which is a very bizarre experience.'

I asked him how he first came to be aware of what he was doing, and able to understand or articulate it. He chuckled and said: 'To be honest I would love to give myself credit for this, but I saw a therapist. I relied on outside help to see this and I frankly don't think anyone can really spot these behaviours without some outside help. I think there is still a bit of stigma about going to therapy, particularly in your personal life, and it seems like that is something you only do at the end of a relationship. That seems like a mistake to me.

'There are things you can't really see about your relationship when you are in it – even your partner can't see – and people outside can spot things really quickly. In my case, when I came to see a therapist, I was pretty readily diagnosable. If I saw someone like me from two years ago, I could probably size them up pretty fast.

'At the time, it feels like your world is ending, so subjectively you are going through something pretty challenging. But someone could quickly diagnose you and very quickly give you some tools to help you correct course. And the more you can do that in the moment, the better, because you stave off your relationship turning even more sour than it might already be. I think that couples should just go to therapists every six months. It's like going to the dentist, you should go to your therapist and just talk about what is going on in your relationship.

'I have seen a bunch of therapists in the last two years … and the first one I saw reacted to my life story with complete shock and awe. She said, "I can't believe your childhood was like this." No one had ever responded in that way because on the surface of it, it seemed fine: I went to school, I was clothed, I was fed, there was no sexual abuse. But there was a lot of what I think therapists would call emotional abuse.* It is a heavy charge to level on your parents but that's how therapists would see it.

'The first therapist was really horrified and I think that you need someone in your life to point out the things that aren't really working, to help you deal with these things. You won't really ever know what is motivating you unless you explore it in depth with someone else.'

Can you change your attachment style?

Our internal models are quite robust and affect our relationships for the rest of our lives. This is largely because they are subconscious, because we lay the foundations of our internal models well before we have any language skills or specific memories, and because, in the case of insecure attachment, they are very powerful defence mechanisms.

I appreciate that doesn't sound very optimistic – but there is some scope for change, and people can move from being insecure to secure. Studies have discovered numerous adults whose lives predict an insecure attachment style, but who are, in fact, secure. Psychologists call this 'earned secure attachment', and you can 'earn security' in a number of ways.

---

* Emotional abuse is an attack on someone's personality and esteem: it can involve putting them down, making them feel like they're going mad and blaming them for the abuse, or humiliating them in front of others. It can also include bullying or control. I talk more about emotional abuse in Chapter 13.

One way is to find someone secure and settle down with them. But insecure attachment systems don't always make this easy. Another option is to find a therapist – but for the therapy to work, the therapist has to *matter*, and insecure attachment systems don't make this easy either. An avoidant person, for example, might subconsciously try to find ways to minimise the therapist's importance. They might see themselves as strong and independent, and the therapist as weak and dependent. Or they might try and stop therapy altogether – which is essentially what they are doing in romantic relationships when they decide to break up when things get intimate.

Assuming the therapist is able to create a secure relationship, their role is then to help their client observe and understand any insecure thoughts and behaviours, which they may not have noticed before. Their role is also to try and help them develop emotional awareness and learn to regulate their emotions. Avoidant people aren't usually very good at this because they can't get in touch with their emotions or bodily sensations, and anxious or unresolved people also struggle because they can't control or regulate their emotions.*

In order to understand the power that attachment styles can have on our emotional lives, as well as how to try and overcome them, we need to take a few moments to explore the structure of the brain.

The brain in attachment

As information from the external world gushes in through our ears, eyes, nose and skin, it meets in the thalamus. The thalamus then acts as a 'cook', stirring this together with

* I'm talking in absolutes, but as I mentioned at the start of this chapter, people will be somewhere on the avoidant and somewhere on the anxious scale. It's not simply a case of being anxious or not.

internal ingredients like expectations and judgements, and comes up with a sense of what is happening to us. This information is then sent to two places, up to the 'rational brain', where it reaches our conscious awareness, and down to two little almond-shaped structures in the limbic system, called the amygdala. This second pathway is incredibly fast, several milliseconds faster than the path to the rational brain.*

The limbic system monitors danger, judges what's enjoyable or scary, and what is or isn't key to survival. It assesses information in a more global way than the rational brain, jumping to conclusions based on rough similarities – whereas the rational brain sorts through a complicated set of options. Its main job is to look out for our well-being and it communicates with us by releasing hormones when it senses danger … or a promising partner.

The amygdala is often called the brain's 'smoke detector', because its main job is to identify whether the information it receives is relevant to our survival. And as it processes information from the thalamus (the 'cook') faster than the rational brain does, it decides whether something is life-threatening before we are even consciously aware that there is any danger. So, before we realise what's going on, our body might already have been kicked into action.

The amygdala doesn't stop and think; it just gets you ready to fight back or flee. And it remembers experiences as emotional memories, which can bias the way we experience things in the present. So, someone who had been hit by a car might find themselves having a strong

---

* Just to warn you – this is a huge over-simplification of the way the brain works. By 'rational brain', I mean the top layer of the brain – the neocortex. We share this outer layer with other animals, but it's much thicker in humans. The frontal lobes make up most of the neocortex and only start developing in the second year of life. Amazingly, the prefrontal cortex isn't fully developed until we turn 25.

physical reaction if a loud vehicle came close to them. But as long as you are not *too* upset, your rational brain* can step in and restore balance by telling you it's a false alarm and stopping the stress response.

It's not just the rational brain that tells the amygdala to calm down – there's a part of the limbic system that also does this, the hippocampus. While secure relationships essentially teach a child's developing brain (including the hippocampus) to calm down the amygdala, trauma in parental relationships can shut down the hippocampus temporarily or inhibit its development. This leaves the amygdala free to be over-vigilant, sensing fires when there's barely a puff of smoke.

In other words, as a result of how and when the brain is built, what we learn to expect plays a huge role in what we actually see. And if, as a result of insecure attachment experiences, what we expect is that people won't be there for us when we need them, this is likely to be what we'll experience. It's a self-fulfilling prophecy.

One more thing about the hippocampus: it doesn't just play a role in calming down the amygdala. It also helps register emotional memories as specific to a place, person and time. But it doesn't come online until the second or third year of life – and we start developing our attachment systems and internal working models within the first. That means that these very early memories are subconscious and often accessible only through physical sensations, feelings or impulses. They also tend to be globalised – because the hippocampus doesn't ground them in a specific place or time. This means that they can be disproportionally powerful.

* Specifically, your frontal lobes.

Earning security

The reason for my brief tour of the brain is to show that 'earning security' literally means rewiring the brain, often at its most basic, pre-language survival level. It can be slow and difficult, but it can be done. One of the most important things for people with insecure attachment styles to learn is how to experience, and regulate, their emotions.

Highly anxious people (as well as those with unresolved trauma) tend to be very sensitive to threats. That's because, as well as having an overactive amygdala, they also have an underactive hippocampus and rational brain, which in secure people work to tell the amygdala to calm down. Highly avoidant people, on the other hand, can't access their emotions – instead they repress them in order to have a sense of invulnerability and independence. And in most cases, they have no idea they're doing it.

So anxious people need to learn how to regulate their emotions, and work on their sense of helplessness and fear of being alone, whereas avoidant people need to get in touch with their emotions and explore their negative view of intimacy.

Developing a healthier relationship with emotions is particularly important for people who've suffered trauma, as they often bury their painful memories. Instead of burying them, they need to learn how to observe and accept the gut-wrenching sensations that register misery and humiliation. The price of *not* doing this, and of ignoring or distorting what their bodies try to tell them, is to deny themselves the ability to know what's harmful and, just as bad, what's safe or nourishing. It also means that they are unable to teach their bodies that the danger has passed and allow themselves to live in the present.

Learning to experience emotions that he'd unknowingly repressed was one of the things that Alexandru worked through with his therapist. 'The first thing that therapy really gave me was awareness, which I think is pretty invaluable. One therapist asked me to read a list of emotions. This sounds really silly because they were all very basic emotions: fear, anger, resentment – and every emotion was broken down by intensity. So, anger initially, then resentment, then hatred, then dislike. I had to live life trying to name whenever I felt an emotion and that was very helpful as a first step. That might not sound like much of a tool, but for me it was pretty revolutionary.'

Learning to label emotions might sound incredibly simple, but research suggests that it's very effective. When people were shown a variety of facial expressions, the area of their brain known to be involved in fear (the amygdala) was much less activated when they were asked to label the emotion of the person photographed.

'It is hard to change,' said Alexandru, 'but I think that you have got to start small. Becoming aware is the first thing. It's very important that therapists help you explore situations that trigger your insecurity and really pick it apart. So, the next time, you stick with the discomfort for a bit and, for avoidant people, the next stage is feeling the damn emotion. You can say to yourself, "I am feeling fine about this," when actually there is an emotion that you are suppressing, you are avoiding completely.

'Hopefully you get to the point where you can actually decide whether these are emotions that you would act on or not and you can divorce yourself from the automated, pre-determined or compulsive behaviour.'

Another important aspect of earning security is to be reflective.* It helps us to have an idea what other people believe, feel, want, hope and plan, as well as what they're

* Psychologists prefer terms like 'reflective function' or 'mentalising', but in the interests of simplicity, and as they are relatively similar, I'm going to merge the two and talk about 'reflection' or 'being reflective'.

trying to hide – and this puts us in a much better position to understand and predict their behaviour. It also allows us not to take things personally – for example, by thinking, 'They were unloving … but I am not unlovable,' rather than, 'They were unloving because I am unlovable.'

Psychologist Peter Fonagy and his colleagues provide some examples of what different levels of reflection look like in practice. On being asked, 'Why did your parents behave that way?' people who aren't reflective might just say, 'I don't know' – giving no indication that they thought about their own actions or the motives of others. Someone who was very reflective, on the other hand, would show a good understanding of other people's motivations, as well as their own. They might say, 'It was very difficult for my mum to like me. My father adored me so much, it would have been superhuman of her not to feel jealous.'

The evidence suggests that reflection can have a huge impact on attachment. Research looking at women who had experienced a very deprived upbringing found that each one who was highly reflective went on to have a secure child, whereas of those on the other end of the scale, only 6 per cent did. In other words, mothers who were very reflective were able to prevent insecure attachment from passing to their child.

There's no quick fix for attachment insecurity. Attachment styles are robust, but they're not unchangeable. Learning to reflect (both on your own mental states and those of others), to excavate and process any trauma you have suffered, and to regulate your emotions can trump your personal history – even when that history is horrendous. In doing so, you can potentially transform the way you relate to others – and yourself.

CHAPTER FOUR

What I Want More Than Anything

'My view of love has changed a lot as I have got older. Now, as someone in their 60s, I don't think romantic or wildly sexual love is as important as companionship. What I want more than anything out of this, my last relationship, is kindness. That's all I want to give and that's all I want to receive.' I met Lynn in Devon. She had short, choppy blonde hair and wore chunky black glasses.

'It was a slow realisation. It came largely in the death throes of my last major relationship, which lasted for 17 years. That was probably five years too long, maybe 10. It was only when we broke up that I was able to identify what I wanted – which was kindness. And Sarah was very unkind.

'I was so afraid of being alone and living alone. I'm afraid I might have ended up with her forever, but I was told to go. Although that felt like the unkindest blow of all, I'm so glad I left.

'Having been together for nearly 17 years, we completely cut contact when we broke up. The first communication was some time afterwards, over a dying cat. She came round to say goodbye and showed a lot of kindness to the cat. But not to me.

'I became completely galvanised. I moved out and started online dating – I just knew I had to do something to counter this unkindness. I also had a good therapist – I think everyone should have one. I had three dates a week for a month or so – I met a lot of frogs. I dined out on stories and my friends gave the people I met fake names, like "The Fascist with a Truncheon" for a policewoman or

"Polly Petroleum" for someone who worked in the oil industry.

'Then Barbara saw my profile ... I remember when we met, I was immediately struck by how kind she was. I wanted to take it very slowly – not in terms of going to bed (we did that the first night), but in terms of commitment ... Eventually we moved in together and we've been together for eight years.'

Kindness

Lynn wasn't the only person I spoke to who thought kindness was important in a partner. Mun Hee, a bespectacled videographer from South Korea, explained that, although his ideal girlfriend did not have to be attractive (the reason being that he didn't think he was), she did have to be kind. And Małgosia, a cleaner from Poland, reflected on the men she met along the way, saying: 'I did meet kind men from time to time, so perhaps it was my mistake not to marry them. But then again, I don't think I realised at the time that they were kind and I didn't appreciate how important kindness was.'

Having dug up a lot of research throughout this project that was less than uplifting (the unsettling frequency of domestic abuse,* for example), I was comforted to hear that people around the world saw kindness as important. And what I was being told in interviews was supported by an enormous study called the International Mate Selection Project. A mammoth 10,047 people across 37 countries and

---

* For example, in an 11-country study by the World Health Organization, the percentage of women who had suffered physical violence by a (male) partner ranged from 13 per cent in urban Japan to 61 per cent in provincial Peru. The numbers for non-physical abuse are even higher. I explore this in more detail in Chapter 12.

six continents were asked to rank 13 characteristics in order of preference – and 'kindness and understanding' was ranked as most important by both men and women. This was followed by intelligence, an exciting personality and health (in that order).*

I like ranking because it forces people to make choices: is kindness more important than intelligence? And where does health fit in? You have to choose. You can't just avoid prioritising them by giving them each the same score. What I love about this mega study is that people in 37 countries put kindness and understanding above everything else in a romantic partner. If that doesn't give you faith in humanity, I don't know what will.

It's also worth pointing out that the instructions were 'sex-neutral' – so 'physically attractive' was used instead of 'handsome' and people were asked to consider what they would like in a 'mate' rather than a partner of a specific gender. This is important because there isn't much research on the preferences of the LGBTQ+ community.† Although the International Mate Selection Project wasn't specifically looking at LGBTQ+ preferences, at least the questionnaire didn't exclude them.

In another study, women were shown photos of men together with (fictitious) descriptions some of which included phrases such as, 'I enjoy helping people' or 'I

---

* One of the limits of this mega study was that the characteristics originated in the US – so there may well be more that were not measured but are very important to other cultures. That said, the study did measure virginity and housekeeping … People in China, India, Indonesia, Iran and Taiwan thought this was very important, whereas people in Scandinavia seemed to think it was pretty irrelevant.

† Just to prove I'm not exaggerating: during 1980–93, fewer than 1 per cent of articles in the *Journal of Social and Personal Relationships* looked at sexual orientation in any way. This was even less (fewer than 0.2 per cent) for the *Journal of Marriage and the Family*.

volunteer at a food shelter.' They were also shown (fictitious) emails about men, for example one mentioned that a man played the guitar, whereas the other said that he played the guitar at a children's hospital. In both cases, women preferred altruistic men.

There is, however, some limited evidence to suggest that kindness isn't always a priority. In one study, people were given a 'budget' that they could 'spend' on qualities in a potential partner. Although kindness was important when people had plenty of spending power, when given a low budget kindness came close to being a necessity – but it was ultimately viewed as a luxury. Given that this was a much smaller study than the International Mate Selection Project, and kindness has prevailed in others, I'm going to dismiss this one and cling desperately to the hope that most people do see kindness as a necessity.

Emotional stability and maturity

As well as having to rank 13 characteristics in order of preference, the people who took part in the International Mate Selection Project (all 10,047 of them) were asked to *rate* 18 characteristics on a scale of 0–3. Just to be clear, this means that, although people had to *score* how important each characteristic was, they weren't being forced to prioritise them (they could give everything a '3', for example).

After 'mutual attraction or love', the top three on average were: 'dependable character', 'emotional stability and maturity' and 'pleasing disposition'. And emotional stability and dependability reminded me of our old friend attachment theory.

'I didn't know about attachment theory when I was doing my tick list and looking for my first husband. But maybe if I'd known about his upbringing, I would have realised that we were not going to be very good together. Maybe you should try and see what someone's attachment style is like on their first date!' Although Diane* said this with a wry smile, there was an element of truth to what she was saying. If you want a partner who is comfortable with commitment and able to process and regulate their emotions, it might be a good idea to be mindful of their attachment style – as well as your own.†

Wealth

'In Chinese culture, how much money someone makes is important when looking for a romantic partner.' I met Tammy, a secondary school teacher, in China. We spoke in an empty classroom filled with whiteboards, grey desks and plastic chairs, with large windows overlooking the sprawling city of Chengdu.

'In some cultures, it might be rude to ask how much someone earns, but in Chinese culture, it is not a secret. Chinese people are very straight talking – they will say, "So you're a lawyer? How much do you make a month?" Say you wanted to be my boyfriend, my family would ask you, "What is your salary? Do you have a house? Do you have a car?"

'It is definitely one of the most important things for a lot of Chinese families. I won't say that money isn't important for me, but before I got married, I used to think, "My future husband doesn't have to have a house, he doesn't have to have a car – but I would like him to have the ability

---

*You met Diane in Chapter 3.

† That's not to say that attachment is the only factor in commitment and emotion, that it should be a deal-breaker, or that you can establish what someone's attachment style is simply by looking at their upbringing. Romantic relationships, trauma and loss can also impact attachment style – as well as a real desire to change, going for therapy and developing your reflectiveness.

to buy things with me one day, if that is what we decide we want to do." Mine is not a traditional Chinese view.'

It's not just certain cultures that highly value money in a partner – many academics argue that women do the same. In the International Mate Selection Project, women across all continents, races, political systems and religious groups valued good financial prospects more than men did.*

Academics offer various theories to explain this gender difference. Evolutionary psychologists argue that it's only natural that women have different preferences to men as they invest much more in reproduction. Much, much more. David Buss explains that men produce millions of sperm (apparently these are replenished at a rate of approximately 12 million per hour!?), whereas women have a limited supply of eggs (releasing around 400 before they reach menopause). After what might be no more than a few short moments, a woman can be left with an energy-consuming nine-month investment, often followed by breastfeeding, which, in some cultures, lasts for years.

Thanks to this massive difference in outlay, Buss argues that women and men value different things: women look for good financial prospects because they are helpful in raising children, and men look for physical attractiveness because it implies fertility and health. (Though that doesn't explain why research suggests gay men also look for physical attractiveness in a partner …†)

---

* The desire for wealth is not limited to humans. Just before the start of the breeding season, male shrikes in the Negev Desert of Israel collect useful objects (like feathers and bits of cloth) and prey (like snails) and impale them onto thorns. The females then come round and check out what's on offer before mating with the males with the largest bounty. When a researcher arbitrarily moved parts of one male's collection to another's, the poor guy lost out because the female chose the shrike that had more.

† I discuss this in more detail shortly …

Other psychologists argue either that there are no gender differences or, if there are, that they can be explained in other ways: women might value wealth, not because it is helpful in raising children, but because they live in patriarchal societies where women often don't work and men hold considerably more power. And there is evidence to back this up. In countries with more gender equality, there is less of a difference between men and women when it comes to looking for cash.

Even though women in the International Mate Selection Project valued good financial prospects more than men, it didn't actually make the top eight things they were looking for in terms of what they rated. And as for what they ranked to be important – 'good earning capacity' came after 'kindness and understanding', as well as 'intelligence', 'exciting personality', 'health', 'easy-going', 'creative and artistic', 'physically attractive' and 'wants children'. So, women are hardly the money-grabbers some evolutionary psychologists would like to portray them as.

'If I had to give advice to those looking for love, I could be really cynical and horrible and say go for the money. But I absolutely didn't do that.' I met Rochelle in an empty train carriage travelling through Cornwall, England. She was in her 60s. We got so carried away talking that we did not realise the train had arrived at its destination and we were thrown off by the train driver.

'One of the guys that I went with – he's dead now – was a college lecturer, and he said, "Marry me for my pension." It would have come in very handy, thank you very much, but I didn't. I don't think the money really matters. After my divorce, I went out with a guy who had a beautiful boat and a big expensive car, but he was totally and utterly boring. You know, the thought of being with someone wealthy is great, but boring is bad.'

Małgosia from Poland explained: 'It doesn't really matter where they come from or what they do. Money is the least important consideration. I don't want a man who has absolutely no money whatsoever because if there is a problem, money helps. But in this day and age, people seem to think that money is the most important thing. They don't see kind men and try and keep hold of them – instead they look for men with money. It's not just in romantic relationships – people seem to be choosing money over health and contact with other people.

'What I find so strange about this is that people are essentially collecting pieces of paper. They're not going to be able to talk to these pieces of paper. They'll be alone with four walls and lots of paper but no one will like them or talk to them. And surely talking to people is the most important thing in life? What will they do if something goes wrong? They won't have any relationships. Will the pieces of paper or the four walls give them advice when they need it?'

Attractiveness

'In my fantasy, he doesn't have to have money.' Joanna was 14, had hair down to her knees and wore an oversized t-shirt with leggings. She was from a village in the East Midlands, England. 'I usually picture somebody attractive, but that's because I can picture it. It's not the main deal – attractiveness actually only comes in a little bit because personality totally outweighs that.'

Like most females, Joanna didn't see physical attractiveness as key (although she did admit it was nice to have). Males, on the other hand, seemed to think it was vital – guys from South Africa, Italy and Greece, to name a few, commented on its importance. And in the budgeting study I mentioned earlier, men with a low 'budget' spent a lot of it on an attractive partner – something the women in the study didn't do.

Precisely what men find attractive depends on their culture.* In cultures where food shortages are common (such as Uganda, Kenya and parts of Ecuador), men prefer women who are heavier and have more body fat. You might think that the opposite would be true in cultures where there is plenty of food – that men would look for women who are skinny – but these men don't actually prefer thin, they prefer *average*. When men in the US were asked to pick their preferred female figure from a selection of nine bodies that ranged in size, they chose the average female body.

Yet when women were asked to choose their ideal body, as well as what they thought a man's ideal would be, they opted for a figure that was slimmer than average. In other words, women think men go for slimmer women than they actually do. Maybe that's why one online dating study found that women often lie about their weight on their profiles (and men lie about their height).

As I said before, there's not much research on LGBTQ+ preferences, but the limited studies that do exist have found that gay men see physical appearance as more important than women do. When researchers trawled through 800 personal ads in the US (back in the 1980s, when people used to advertise in personal columns), they found that gay men were more likely to ask for physical attractiveness than lesbians or straight women or men.

This is just one study from one country and it relies on what people publicly said they wanted. On that basis, it is

---

* As well as the individual. Adam, who I met in Antarctica, said: 'Maybe certain things about me are not as attractive to the normal woman. And normal women are not necessarily attractive to me – I'm very picky. It's not necessarily a conscious decision, but I think some people are attracted to every pretty face they see and that's just not true with me.'

impossible to generalise. Paul, who I met in Kashgar, China, provided an important reminder of this: 'Historically, gay men have appeared well-groomed, feminine and muscular – and I think that's what society expects. Although some gay men might find physical appearance more important, some won't, in the same way that straight or bisexual men may or may not. I don't think what you find attractive is down to your sexuality.'

There is very little evidence to confirm or deny whether Paul is right, and whether your sexuality impacts what you find attractive, but there's plenty to suggest that your gender does, specifically when it comes to physical attractiveness and age.

Even if men do value attractiveness, studies suggest that they don't like to aim too high – perhaps for fear of rejection. Instead, unless they have reason to believe that someone beautiful is interested in them, they tend to go for someone who is roughly as attractive as they are. In one study, for example, men were shown six women. Of those who were told that *all* six women were attracted to them, 78 per cent went for the most beautiful. When they were told nothing at all, they went for someone less attractive.

Age

Sean, a pro American footballer who I met in the US, told me about friends of his who were 'having a midlife crisis where they buy cars and get with women the same age as their daughters'. To avoid this risk, he advised not getting married young and instead 'getting it all out of your system'. The evidence backs him up: whatever their age, whether gay or straight,* men prefer their partners younger.

* I didn't come across any research exploring the age preferences of bisexuals.

Just how much younger depends on their own age. Psychologists Douglas Kenrick and Richard Keefe found that men in their 30s preferred women who were roughly five years younger, but when they got to their 50s, they preferred them 10–20 years younger. In other words, the older a man, the younger he will want his women to be.

Meanwhile, when researchers trawled through 783 personal ads from a variety of magazines across the US (this time in the 1990s), they found gay men were very similar to straight men in their preference for someone younger. In their 20s, both gay and straight men were happy with someone five years either side of their own age. But, as they aged, they tended to prefer younger and younger partners. When they were 50 years old or more, straight men wanted someone 5–15 years younger, with gay men preferring even younger partners (5–20 years).

When Christian Rudder, co-founder of large online dating site OkCupid, examined the data of straight male users aged between 20 and 50 years old, he discovered that, no matter how old they were, they rated women in their early 20s as most attractive.* In other words, men's idea of what is attractive doesn't age with them – at 50, they fancy who they did when they were 20.†

If you examine who they message most, it looks like this: from 22 onwards, men start messaging women younger than themselves. The age difference increases as

* Although these stats refer to OkCupid data, these findings hold for all major online dating sites Rudder looked at. I should point out that Rudder analysed the data of white men aged 20–50. He did this because race had a massive impact on results and he wanted to isolate age as a factor. With the exception of 45-year-old men – who for some reason prefer 24-year-olds – the rest prefer 20 to 23-year-olds.

† Well, in fairness, the data suggests that at 50 he thinks 22-year-olds are the hottest, whereas at 20 he thinks 20-year-olds are. So in 30 years of ageing, the age of the women he fancies goes up by all of two years.

they get older until, at 36, they seem to decide that they really shouldn't be hounding anyone in their 20s any more, and from 36–40, they mostly get in touch with 30-year-olds. Then, at 41, they decide that even that is too young and start mostly messaging 35-year-olds. By 45, they have decided that 35 is too young, and jump to messaging 40-year-olds. Visually, it looks like this:

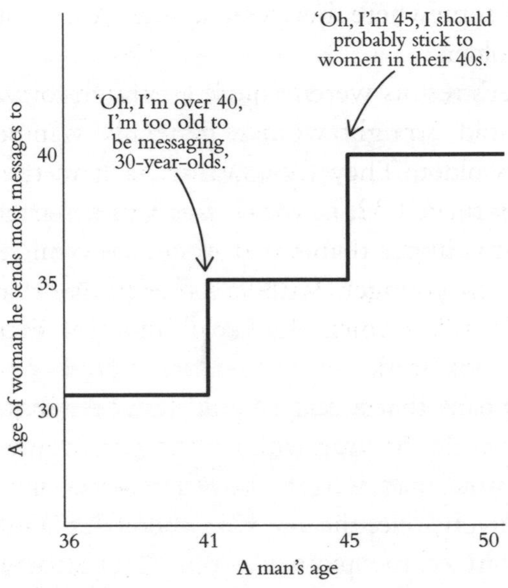

What's particularly interesting about the OkCupid findings is that if you only looked at what men *said* they wanted, you would come up with a very different picture – men *say* they want ages increasing in line with theirs. But who they actually get in touch with and rate as attractive doesn't fit that story. Rudder wasn't looking at the results of an artificial experiment. Men weren't being asked face-to-face, there was no incentive for them to lie – in fact, there was an incentive to be honest because getting it wrong meant that they would be sent profiles of women they didn't actually fancy. Yet they still said that they wanted women older than the women they *actually* went for.

The situation isn't as clear cut when it comes to women. In the study of 783 personal ads I mentioned earlier, researchers found that as lesbians got older they wanted increasingly younger partners (though not quite as young as men would like). Straight women, on the other hand, didn't change with age – instead, they consistently looked for men somewhere between a few years younger and 10 years older.

Rudder's results weren't quite as straightforward. Up to 29-years-old, straight women generally wanted a man a few years older. They then wanted a man the same age until they turned 32, at which point they started to prefer men a bit younger than them. Once they turned 40, that became a *lot* younger. Rudder's theory was that women's tastes hit a wall (or men's looks fall off a cliff, depending on how you look at it).

This means that a 32-year-old female will sign up, set her filters at 28–55 years old, and start to browse. Whereas a 35-year-old man will set his filters at 24–40, but rarely contact anyone over the age of 29. Of course, online dating is only part of the picture – plenty of people meet via friends, the workplace, etc. – and maybe those people go for others who are similar in age to them. But given the size of the sample and the variety of the demographics of the people involved, I don't believe that it's only male online daters who are looking for a younger partner.

The older woman

There is, however, one exception to men's tendency to go for someone younger. Rudder found that 40-somethings were less likely than men half their age to go for a woman in her 40s (who you might know as a 'cougar'). That means women in their 40s might have better luck going for someone in their 20s than someone the same age. On a Friday night in Glasgow city centre, I met one such couple.

Ayo, who was originally from Kenya, was standing outside an abandoned pub eating pungent takeaway from a white polystyrene container when I asked him for an interview.

'Age is just a number, it doesn't matter. One person could be 40, and another could be 20 – as long as you click and share the same values, age is nothing. The difference between your age is nothing.

'Put it this way, I prefer an older woman because they are more mature. I'm 23 and my girlfriend is 41. Let's say I went for a girl my age, they are more likely to be wanting to be out there and fucking about. Sometimes you want a proper relationship and not a runabout. A younger woman could be nothing more than a one-night stand.

'You can't really say that you will end up with an older woman, a younger woman – as long as they are legal and you both have the same views and values, then I think you've got nothing to lose. Get it on, get it going, hit the road, see where the road takes you. That's my opinion.'

Humour

I met Jinny in a bustling cafe in Spain. She was olive-skinned, brunette and hoping to meet a someone to settle down with. 'I've changed in terms of what I look for now. Looks don't bother me at all any more. My main criteria is someone who makes me laugh and makes me feel comfortable because that's enough for me to find them physically attractive. I don't think my parents would have been married as long as they have if they didn't make each other laugh. I think I need someone to take me down a peg or two when I'm being too serious or when I'm worrying, and the best way to do that is with humour.'

Research in Western cultures has found that both sexes think a sense of humour is important – but in different ways. Men prefer people who laugh at their jokes, whereas women want someone whose jokes they can laugh at (although, surely, the ideal is to have both? I can't imagine that having someone who just laughs at your jokes without ever making any of their own is really living the dream).

In terms of what they find funny, women enjoy flippant one-liners rather than direct pick-up lines. They also seem to enjoy self-deprecating humour (as long as the person making the joke is attractive), whereas men can find this to be a turn-off in a woman. Both sexes prefer positive to negative humour, but when looking for something short term, women seem to be less fussy and don't mind a bit of pessimism. This made me wonder whether someone who just fancied a fling and nothing more could maybe use negative humour to filter out people who wanted something long-term …

Self-awareness

Time and time again, research shows that what people say they go for isn't always what they actually choose.* This suggests a distinct lack of self-awareness, at least when it comes to what we want in a partner – but then self-awareness† is anything but easy. In fact, some argue it's impossible.

Ancient Greek philosopher Aristotle believed that awareness was key to being a good person, but that it was hard to really know yourself – this, he thought, was clear from the way we blame others for doing things that we do (without realising) or claim that we have virtues that we

* I'll explore this in more detail in the next chapter.

† It might also suggest that they are conscious of what others will think of what they choose and are trying to choose in line with those expectations.

don't. Meanwhile, German philosopher Friedrich Nietzsche thought we could never really understand our actions, let alone our motives, because they're too complicated and opaque.

Whether difficult or impossible to fully understand ourselves, some of the people I interviewed saw at least a degree of self-awareness as essential in a partner. Alexandru, who was single and from Romania,* explained, 'I look for people who have a glimmer of self-awareness, someone who has given some thought to their interior world in some way. You would be surprised – virtually no one has. They don't want to be there the moment you even ask them. It's not like I go to dates and offload about the stuff that is happening to me (well, maybe a bit). I ask them about what is going on with them, how they relate to their past, what they are doing now, what they want to do or literally anything about themselves, and they will give me a very safe, socially acceptable answer.

'I have been on dates with very amazing women, who are doing something really incredible. They don't need to have low self-esteem, but a lot of them can't express themselves beyond what they have done. And that is fine, I don't want to knock people's achievements, but I am looking for someone who gets what I am saying.

'I have experimented with this, and frankly the more emotional and more honest my response, the more they just shut down. They just don't really believe me. If you give the cookie-cutter "I am busy right now" response, that somehow is internalised more easily. But if you say, "One of the reasons I don't feel comfortable seeing you any more is because I like you and it's scaring the shit out of me," they won't say it, but you can tell with their body language they don't buy it.

---

*You met Alexandru in Chapter 3.

'I could play the game to a certain extent but that becomes unsatisfying and I don't want to just date people for the hell of it. The point is to feel connected and to fall in love with someone. I think that love is admiration and I am finding it hard to admire people who don't reflect at the moment.'[*]

Big decisions

It's one thing discussing what other people look for in a partner, but exploring their decisions is not the same as actively questioning what *you* want.[†]

Of course, what you look for isn't completely rational and is influenced by all sorts of things that you are not aware of. As a result, you might find yourself attracted to someone without quite being able to put your finger on why (or rather, you'll probably think you can when you can't). Biology plays a role. For example, when you meet someone, you sniff out their immune system, preferring systems that complement yours[‡] and what women look for in a partner, as well as how they think and behave, is hugely influenced by where they are in their menstrual cycle.[§]

---

[*] If attachment researchers are right, this sort of reflection will be massively important in people who have an insecure attachment style – without it, they may not 'earn' security.

[†] Assuming you want one, that is. In Chapter 6 I explore why you might not want a partner.

[‡] The theory being that if we reproduce with someone who has a complementary immune system to our own, our children will have a wider set of immune responses.

[§] I explore this and other biological processes in more detail in Chapter 5.

The 'they'

The hunt for love can be just as influenced, if not more, by the outside world – whether that's the culture we find ourselves in, social norms, friends or family.

I met Farsideh in a cave in southern Laos. She was glamorous, Persian–American and offered to share her torch with me when mine started to run out of battery. 'Sometimes friends of mine talk about meeting someone so much that it drags me down with them. That's when I start dating people I know I don't even like.

'A lot of them say they think outside the box, but when it comes to love and relationships they are so firmly in the box that they don't even realise there is a box. I have so many friends who are otherwise very forward-thinking and empowered, but it's never occurred to them that they could create the life that they want for themselves.

'A lot of people think that unless they achieve what society expects of them when it comes to romantic relationships and having children, it's less valid or not right. They won't admit it, but they think it. They don't want to be the person who has to explain why they are 43 and single. But why is it anyone else's business?'

In their own distinctive ways, philosophers Kierkegaard, Nietzsche, Heidegger and Jaspers all wrote about the anonymous mass or the 'crowd' who influence and even dominate the way we think and behave. German philosopher Martin Heidegger used a word often translated as the 'they' to describe it.* The 'they' isn't a particular group of people or a measurable thing; it's a powerful authority

* There's a lot of controversy about what exactly is the right translation, because the 'they' suggests that we are somehow separate from it, when it's not that straightforward and we are actually part of it. For that reason, some philosophers argue that a better translation is 'anyone' or 'one' or 'people'.

you can't pin down. It's nebulous and vague, but, without our realising, it has a huge impact on the way we think and behave.

Heidegger believed that it's all too easy to follow the 'they' – to enjoy ourselves in the way that the 'they' enjoy themselves, to be shocked by what the 'they' finds shocking. But he thought that by following the 'they', we avoided making our own choices and, ultimately, living our own life.

Another way of explaining it is to think of 'what one does' – or at least that was American philosopher Hubert Dreyfus' interpretation. A classic example is how far away you stand from someone when speaking to them. Of course, the appropriate distance changes depending on where you are and how well you know the person you are talking to, but you usually have an idea of what's acceptable and you comply. If someone you had only just met stood a little too close to you, you would notice, perhaps even disapprove. It's just not 'what one does'.

Does it really matter if someone stands a bit too close when they speak to you? What's the reason for having an appropriate distance anyway? Heidegger explained that 'what one does' doesn't need any justification: we simply absorb what we should or shouldn't do and obey. And it's not conscious – you probably don't stop and calculate how far you are from someone before starting to speak (or at least I don't). You just conform.

When it comes to what we look for in a partner, the 'they' can be very powerful.* Lynn (who you met earlier in this chapter) was so influenced by 'what one does' that she spent years dating men even though she was only attracted to women.

* Of course, the 'they' don't just influence what we look for – the 'they' also have a huge impact on how and why we look for someone.

'The first person I ever loved was a nun. I was only 12 and she was my teacher. I graduated from that to boys – I did boys until I was 33. I always knew that something wasn't right for me in those relationships. We're talking the 1960s and 70s – I'm a nice middle-class girl and it was understood that I would meet a nice boy and have some babies. I was stuck in that expectation myself. So I pushed all this other 'naughty stuff' away.

'I lived with one bloke for four years. It was in that time that I thought, "Well this is really lovely and he is really kind," but when it was bedtime, I was thinking about women. I thought, "I can't carry on with that pretence." When we split, I dabbled a bit with other men, but then I met the woman who became my first girlfriend.'

Lynn may not be the only person who has pretended to be straight to conform. When Seth Stephens-Davidowitz analysed anonymous Google search results across the US, he found that searches questioning a husband's sexuality (e.g. 'Is my husband gay?') were far more frequent in the least tolerant regions. In 21 of the 25 states where this question was most frequently asked, support for gay marriage was lower than the national average.*

Sexuality wasn't the only topic to come up. Chess was from New York. She had an angular haircut, a cable-knit jumper and large earrings.

'My grandmother would say, "Get married because otherwise he could leave you," even though she was a politician all her life and I would never have expected her to say something like that. Even my mother, who I would think of as a right-on woman – strong and independent and all that – would still say, "Marry well." This information is passed on between generations and it persists – "Marry wealthy and you won't need to work, even though you'll

* The states with the highest percentage of women asking this question were South Carolina and Louisiana.

have a miserable life. You'll have no self-fulfilment, but you'll have cash."'

Despite having decided to ignore the 'they' and stick to what she felt was right for her, Barbara, who I met by the seaside in the south east of England, still felt tossed around by its power. 'People generally expect you to get married and have kids. I don't know where the expectation comes from. People just expect you to do what they want, maybe. But my reaction is, "Maybe I don't want what you want." It's almost like they're offended because I'm rejecting their choices.

'But then sometimes I think, "Maybe we should get married," because that's what the norm is and I feel like we're doing something that's a bit weird. Or I think, "If we just did it, life would be a bit easier because then I wouldn't have to deal with people's reactions." But then I think, "No – because they're not my values, and they're not my boyfriend's either. I can't just change because of other people."

'A lot of people follow what is expected of them without thinking about what their own values are. I'm sure 10 or 20 years later, they'll stop and wonder, "Whose life am I living? What have I done? Where have I ended up?"'

Shaping each other

So what if you make the wrong decision? So what if you do what is expected of you or go out with someone who doesn't share your values? With the proliferation of online dating, surely it's easy to swipe right and find someone else if it doesn't work out?

Apart from devouring your time, intimate relationships change you, for better or worse. A well-chosen partner can help you reflect on what's important and stick to your goals, keep you laughing when you're taking yourself too seriously and keep you in line when you're being judgemental or unkind. A poorly chosen partner, on the other hand, can grind you down until you are barely visible.

Paul, who I met on the Silk Road in China, put this brilliantly. We met each other standing next to a stall selling dried snake skins and animal heads in Kashgar.* He was a teacher in Shanghai and was travelling during the school holidays.

'When I met my boyfriend, I was attracted to the whole package. Most of the time, with everyone else, it didn't click, but it did with John. There was a spark, a mutual physical attraction, great conversation. Then when you start getting deeper, it was also about our social background, financial stability, likes, that type of thing.

'I don't think you have to like the same things. Otherwise you end up going out with a clone of yourself. But I think you get more similar over time. You rub off on each other. I guess when you meet someone you get on with, you're like two cogs working together. You may not necessarily fit, but over time you shape each other.'

I asked Paul how John had shaped him and he thought for a moment. 'There are obvious things like feeling more love and having that support, which means I feel less anxious, more settled and more stable. And I am a bit more self-aware, in a good way, a bit more confident in knowing who I am – especially as a gay man, it has helped me come to terms in accepting that.'

He paused.

'I'm trying to think of negative things ... but I can only come up with positive. Which is good because we've been going out for 13 years now. We got together in my 20s, when my character was still being defined. Well, it's

* I had hoped to talk to more locals, but my Mandarin wasn't good enough and I didn't speak Arabic (still don't). And no one spoke a word of English in Kashgar – so much so that it didn't really matter what I ordered in a restaurant, I was always given a spicy bowl of meat with a hunk of stale bread. Or perhaps that had less to do with my linguistic skills and more to do with their culinary repertoire.

probably being defined all the time as life goes on and on and on.

'My close friends also shape me. One of them has helped me have a more positive, can-do outlook. I think I'm borderline between being an introvert and an extrovert – he is an extrovert and he raises people up with him. I think that's helped me become more outgoing myself, more of a protagonist. He's also forward-thinking and energetic, and that's rubbed off on me too.

'University wasn't a good time for me, I wasn't particularly happy so I felt quite anxious after that. And the friends I made in Shanghai helped me to feel calmer and happier in myself, which is funny because another friend I have is great, he's a good laugh, but he's also quite anxious, and it rubs off on me and leaves me feeling more anxious myself.'

Maurice, a 95-year-old poet who had been widowed after more than 65 years of marriage, said: 'I suppose my wife did change me. I wouldn't admit it perhaps at the time. But, reluctantly, yes, I think she did. I suppose she changed me gradually from an early stage. She opened me up a little bit to the rest of the world, put it that way.

'I think you might tend to adjust rather than call it change. You probably make certain concessions along the way and then adjust to living with it. I don't think you should try to adjust yourself too much, unless you realise that you were wrong or unkind, in which case you should. But with some of the more miniscule matters, who cares?'

There's very little long-term research investigating this idea of partners shaping each other, but what there is suggests that Paul and Maurice were right. In 1956, psychologist K. Warner Schaie began the Seattle Longitudinal Study with the aim of exploring people's psychological development as they age. Every seven years since, Warner Schaie and colleagues have followed up with the people who took part (as well as recruiting more).[*] In total, almost 6,000 people aged 22–101 have participated and the study has provided fascinating results.

The study followed 178 married couples for up to 35 years, measuring all sorts of data, including intelligence, education and age.[†] On top of that, people were asked to rate how happy they would say their lives had been up until that point, on a scale of one (very happy) to five (very unhappy). When researchers analysed the masses of data, the results were pretty clear – over time, partners became more similar in terms of how happy they were. But that's not all. Over a shorter period (up to 14 years), researchers found that married couples became more similar in both vocabulary and intellectual ability.[‡]

The results suggest that who you choose as a partner can have a big impact on you – something philosophers have long recognised. Aristotle believed that your partner could have such an immense impact that the wrong one could thwart you in becoming a good person.[§]

A good partner, on the other hand, can do the opposite. It makes intuitive sense: it's easier to figure out how you want to be good, or what being good means for you, if

---

[*] Although people dropped out along the way, as is often the case with this sort of study.

[†] The study followed 178 couples for up to 35 years and 169 for up to 21 years.

[‡] Specifically, vocabulary that they would understand, rather than actively use. The analysis relating to intellectual ability was carried out in the 1990s, when the study hadn't been going for quite as long – so this is true for the 169 married couples they followed for up to 14 years.

[§] Aristotle was actually talking about friendship here because, like many philosophers, he didn't think much of passion or romantic love. But what he calls 'perfect friendship' looks a lot like what an intimate romantic relationship could (or should) be.

you're going out with someone who's willing to consider the question in the first place. A good partner can help you develop an awareness, not only of your own inner life but also of theirs and, in doing so, Aristotle argues that you ultimately learn more about yourself. He also believed that things are often easier when you do them with someone else – and that includes being good.

You are your decisions

What you look for and who you choose isn't just important because your partner will change you over time: some philosophers argue that the decision itself shapes you. Greek philosopher Pythagoras believed that life was the sum of all of your choices,* and some, like French philosopher Jean-Paul Sartre, go even further. Sartre argued that you're not just choosing your life when you make decisions, you're choosing who you become.

It's a daunting prospect because it suddenly places a huge amount of importance on every choice that you make. But it can also be liberating because every decision becomes a fresh opportunity to shape yourself in a different way. Historically, you may have looked for people with money or looks, but perhaps today you'll prioritise kindness or self-awareness. In doing so, irrespective of whether you end up in a relationship or how they may shape you in the future, you have shaped yourself simply by making that decision.

Of course, choosing a partner isn't just one decision, it's a series of decisions, some bigger (like who to date or whether to formally commit), some smaller (like whether to support them in a creative project or listen when they

* You might remember some of his thoughts on triangles from school.

tell you about their day).[*] And decisions don't have to be active. You can still 'choose' by simply continuing along a path without changing it (or your attitude towards it). Staying in a relationship or continuing to find problems with everyone you meet in order to avoid intimacy or commitment are still choices – and choices that change who you are – even though you might not experience it in that way.

If Sartre, Pythagoras and other existential thinkers are right, then with each decision, big and small, active or not, you are shaping who you are going to be. And if that's the case, perhaps I began this chapter by asking the wrong question. Maybe instead of asking: 'What should you look for in a romantic partner?' the real question is: 'Who do you want to become?'

---

* I explore these more in Chapter 8.

CHAPTER FIVE

Looking for Love

'I have love in my life. I met her at work. In all honesty, I wasn't expecting it. She was the one that chased me: she knew she wanted to be with me almost the moment she saw me, so she was quite bold. She didn't say anything at first, but after about a month, she approached me.' I met Jorge, a gardener, in rural Spain. He was a shy man but agreed to give me five minutes of his time (and no more) to talk about love.

'When we started going out, I was quite reserved and didn't throw myself into it, but she did. After three or four years, I realised that I wanted to be with her long term. Before that, we didn't really cultivate our relationship. She was going through a divorce and she had a small child, so we had to wait for some time before we could focus on our relationship. We were also trying to get to know each other, find a place to live. It was complicated.

'Now we've been together for 20 years and it has been absolutely marvellous. Of course we have had our ups and downs, like all couples – we fight, we argue. The whole world does it. But most of the time our relationship is very good.

'I'm very happy. I'm not interested in other women. What I want is what I have – I don't need anything else. We are completely transparent with each other, there is no deception, we always say it as it is. I told her back then and I tell her now: if she hadn't found me, I don't think I would ever have settled down. I am very lucky.'*

* It was at this point that Jorge reminded me that he was only giving me five minutes of his time – and it had already been six. I thanked him for his time and he got back to mowing the lawn in the afternoon sun.

I hardly met any women who'd done the chasing, nor did I meet many men who'd been chased. As someone who had done the chasing myself (that's how I got together with my now husband), I was keen to know whether Jorge and his wife were outliers or not. Was he really that 'lucky' to meet a woman who was prepared to pursue him? As soon as I delved into the research, I realised he was – it's almost always men who make the first move.

In a 2016 study of single people in the US, only 13 per cent of women had asked a man for his number and only 29 per cent had initiated the first kiss. (Which is odd – why were more than twice as many women brave enough to physically move in for a kiss than ask for a phone number – surely being rejected when you're trying to lean in and lunge is more embarrassing?) Meanwhile, in an online dating study of more than 8,000 men and 6,000 women, sociologist Derek Kreager and colleagues found that women were four times less likely to make the first contact than men.

Perhaps women expect men to do the asking, perhaps they believe it's better to play hard to get, or perhaps it's just not 'what one does'. I suspect fear of rejection plays a role, but that doesn't stop men – and the evidence suggests they're far more likely to be rejected than women.

In a classic study, male and female students approached strangers of the opposite sex and said, 'I have been noticing you around campus. I find you to be very attractive.' They then asked one of three questions: 'Would you go out with me tonight?', 'Would you come over to my apartment tonight?' or 'Would you go to bed with me tonight?'

Half the men (and women) said yes to the first request, but when it came to being invited to their apartment or to stay the night, every single female said no … and a massive 69 per cent of men said yes. What's more, the attractiveness of the person doing the asking had *no impact* on the response.

Of the men who said no, many apologised or gave explanations, like 'I'm sorry, I'm married' or 'I can't tonight, but tomorrow would be fine.' The females, on the other hand, gave no excuses. Instead, they were appalled, saying things like, 'You've got to be kidding?' and 'What's wrong with you?'

Yet, if more women were to pluck up the courage, perhaps they would be pleasantly surprised by the results. In a recent singles study of more than 5,500 people in the US, 95 per cent of men surveyed were pro women lunging or asking for their phone number, and 94 per cent were pleased when a woman was the first to call after a (good) first date. Meanwhile, Kreager and his colleagues found that women who made the first move ended up connecting with people who were more desirable* compared with those who waited for people to get in touch with them.

Kreager's finding is backed up by something called the 'Gale-Shapley' algorithm. In order to explain it to you, I need you to imagine you're at a party with an equal number of single straight men and women, each of whom have a clear idea of who at the party they would like to date. When I say 'clear idea', I mean they have an order of preference (with no one holding joint first, second or any other place).

Now, imagine the men go around and ask out the women at the top of their list. The women say, 'Why yes, a date would be lovely' to the person they like the most and, 'Absolutely not, no way, leave me alone' to everyone else. Then the men, some of whom have faced cruel and bitter rejection, try again and ask out the next woman on their list (irrespective of whether someone else has already

---

* Desirability was measured by other people's subjective ratings on the site.

asked her out). She then replies and if she has already said yes to the first man but prefers the second, she simply rejects the first and goes with this one instead.

It doesn't matter how many men and women are at the party, the results are always the same:

1 Everyone finds a partner.
2 Every man will end up with the best woman available to him.
3 When everyone is finally partnered off, no couples could improve their situation by running off with someone else (why would any of the men change things? They've each ended up with their best possible option).
4 Every woman will end up with the *least bad* of all the men available.

When I first learnt about this algorithm, I was flabbergasted by this last point. But maybe it's not that surprising: the men approached who they most wanted to be with and if they got rejected, they moved to the next on their list. They kept going and going until, finally, someone caved in and said yes. So, in taking the initiative and risking rejection, they maximised their chances of getting someone as near to the top as possible.

The women, on the other hand, waited for men to come to them. As a result, they ended up with people lower down on their list, not because they actively chose them, but because they were the ones who came over to ask them out.

Of course, when it comes to finding love, this is completely unrealistic – how many parties do you go to where there's an equal number of men and women, they're all straight, they have a clear order of preference and no one is already taken? Mathematicians have adapted the algorithm to allow both men and women to approach either gender, to have joint preferences, or to leave on their own if they prefer (rather than with one of the people who approached them).

The maths can get quite intense so I'm not going to go into it, but for all the variations, the message remains the same: if you can face rejection, you will get much better results if you are brave and proactive instead of sitting back and waiting for others to come to you.

Box-ticking

'Online dating is an interesting world.' I met Adam in Antarctica. He was tall, slim and always wore the same cerulean blue fleece. I later discovered this was because he had sold all of his belongings and was travelling the world with a rucksack smaller than a gym bag. I interviewed him in the library of the ship we were travelling on, overlooking luminescent icebergs and the occasional wandering albatross.

'I think it's gotten so popular it's trying to change the dating scene in general: guys are less likely to go up to girls in bars and much more likely to find them online. It's much safer from an emotional standpoint. If you go up to a girl in a bar and she's not interested, the rejection is very personal and it's hard not to let it affect you. Whereas online, if they don't return your message, you might not even remember that you sent one. The women get so many messages online, you expect that most will not get back to you. It's easier to think, "Well, maybe she's not checking her messages" or "Maybe she's getting a lot" than to take it personally.

'It's actually very difficult to tell what someone is like from their profile. Usually you can have a pretty good idea within the first five minutes of meeting someone – then you either have to be rude, which I'm not, or hang around until it's polite to leave. Sometimes people are still extremely interesting to talk to, even if you are not interested in going out with them. But not always.

'The site I use asks you survey questions and you can rate how important the answers are – you list the 'acceptable answers' of the person that you might be looking for. So, the question might be, "Do you smoke?" and you can mark the acceptable answers and just how important these answers are. If you really don't want to date a smoker, you would say, "No I don't smoke, and the only acceptable answer is that she doesn't smoke." You go through lots of questions and the website finds matches for you.

'It frustrates me a little bit because some of the questions that I would find important are not on there. It's more to do with values. I'm interested in whether a girl is adventurous or energetic and there's not a whole lot about that. It seems like I attract a lot of vegetarian atheists. I don't know why that is.'

Online dating has become increasingly popular (although that's not hard to achieve given that the percentage of couples in the US who met online in 1990 or before is zero), but it's difficult to say exactly how popular it is. In the US, the How Couples Meet and Stay Together study (HCMST) of more than 4,000 people found that 22 per cent of couples met online, making it the third most likely method of finding love – after meeting through friends, and in bars or restaurants.

Other, admittedly smaller, US studies have yielded different results. One found that only 8 per cent had met someone online and another found that only 19 per cent of single people said they ever used online dating. Though, as with infidelity, there may be an incentive for people to hide the fact that they're doing it as online dating can still raise eyebrows. In a 2015 study, 23 per cent of Americans questioned agreed with the statement that 'people who use online dating are desperate'. Ouch.

What I can tell you is that it's way more popular among same-sex couples, with more than 60 per cent meeting online in 2008 and 2009. Sociologist Michael Rosenfeld, the man behind the HCMST, believes this huge difference in popularity is because, when it comes to dating, the LGBTQ+ community is nearly always in a 'thin market' – one where there aren't many buyers and sellers. And one of the benefits of the internet is that it allows people to extend their market relatively easily.

Although online dating can be a more efficient and precise way of looking for love, it's not without its downsides: the precision of the process allows, or even encourages, people to become more prescriptive than they would otherwise be. Just because you *can* filter out people according to whether they like hiking, smoking or eating meat, doesn't mean you should.

As Adam explained: 'Overall, I have been on maybe 20 or 25 first dates, and two second dates. I don't think I would get a better second date ratio if I was going through friends of friends. Online dating really does let you target the type of person that you are looking for because you can see their profile before you message them. With friends of friends, it's less targeted. At the same time, part of the problem is that people are much pickier with who they will go out with when online dating.

'If you meet a girl out in the wild, she knows less about you so it's more about your personality from the beginning. It's easy to get very specific online in terms of what you're looking for and eliminate people that maybe you shouldn't have eliminated. "She doesn't like hiking, so it's not going to work out," but maybe that's not such a big deal after all.'

The best way of demonstrating the impact of being overly picky is to return to mathematics.* In a particularly optimistic moment, economist Peter Backus decided to focus his numerical skills on dating. Specifically, as a single

* The maths used in this example is based on men meeting women because it makes the calculations easier, but the results can apply to any gender and sexuality.

straight man, he wanted to calculate how many potential women there were for him to date. He did this by breaking down the problem into smaller and smaller chunks until he could make an educated guess.

Backus decided that he wanted to meet someone nearby (so that they could get to know each other more easily and save money on train fares). He also wanted someone close to his own age. He was 31 at the time, so set his age range to 24–34 years old (his reasoning being that he didn't want to fall prey to a voracious cougar, nor feel older than he was by not being able to keep up with a spritely 20-year-old). He worked out:

1 **How many women lived near him** – at the time, he lived in London, so that's about **4 million** women.*
2 **How many were likely to be in the right age range** – around 20 per cent, which left him with **800,000**.
3 **How many were likely to be single** – he estimated 50 per cent, which left him with **400,000**.
4 **How many were likely to have a university degree** – around 26 per cent, leaving him with **104,000**.
5 **How many he was likely to find attractive** – he estimated 5 per cent, taking him down to **5,200**.
6 **How many were likely to find him attractive** – he rather pessimistically guessed 5 per cent, bringing the number down to **260**.
7 **How many he was likely to get along well with** – he guessed 10 per cent, leaving a final total of **26 women**.

* Backus relied on data from the Office for National Statistics in the UK to come up with these numbers.

To be clear, according to Backus, that's 26 women in the entire world that he would date. And, as a result, he concluded that there were more intelligent alien civilisations in the galaxy[*] than potential girlfriends for him.

His odds aren't particularly uplifting, especially if you're currently single and looking for love, but before you reach for the vodka I would argue that he is probably being pessimistic. Either that or he is unreasonably picky. Would he really find 95 per cent of women unattractive? And is a university degree really necessary?[†] By changing the responses to questions 5, 6 and 7 to 20 per cent each time, Backus would suddenly have 32 times more potential girlfriends (specifically, 832). That's nearly 1,000 across one city (rather than the entire world). Now, that's a lot less depressing.

Backus' calculations help to demonstrate an obvious but important point – the pickier you are, the slimmer your chances of finding someone.[‡] Of course, it's OK to have some must-haves – in fact, it's probably advisable. But not *so* many that you are essentially avoiding any possibility of love (avoidant attachment anyone?).

What's more, there's a lot of evidence to suggest that people are being picky when they don't even know what they are looking for. Some sites almost ignore what people *say* they want – instead, they build algorithms to suggest

---

[*] This is based on the 'Drake equation', developed by Frank Drake at the National Radio Astronomy Observatory in Green Bank, West Virginia. Drake estimated that 10,000 communicative civilisations probably exist in the Milky Way alone.

[†] Backus actually heads up this point by explaining that he knows there are plenty of intelligent people who don't go to university, but that he, like everyone, is allowed to have preferences. (He then asks how many women out there have dated men shorter than they are, to make the point.)

[‡] Maybe Backus reached the same conclusion and decided to be less picky, or perhaps he just beat his own odds – he is now married.

matches based on who people *actually go for* because there is a big difference between the two.

Speed dating research has found similar results. (In case you haven't had the pleasure of going speed dating, you're given less than 10 minutes in which to impress/assess a potential date, before moving on to the next victim.)

In a US study of almost 9,000 speed daters, most of the time what people said they wanted had little to do with what they actually went for during events. In another study conducted by psychologists Eli Finkel and Paul Eastwick, students completed questionnaires before, immediately after and then every three days for a month after attending a speed dating event. When Finkel and Eastwick compared what students said they wanted with what they actually went for, they found that the two were unrelated. In fact, they concluded that people have little insight into what they truly want in a romantic partner.

On top of that, what people *think* is important to them often isn't. For example, a lot of decisions both online and in the wild are driven by looks, but maybe they are not as important as people (particularly men) think.

Online dating site OkCupid launched an app where users couldn't see what potential dates looked like. Instead, they could only see a scrambled photo.* The app was eventually shut down (it turns out people *really* want to see the face of the person they are going to go on a date with), but not before 250,000 people downloaded it and 10,000 used it to meet up.

After each date, the app asked each person how it went. OkCupid co-founder Christian Rudder compared their

* In case you can't picture a scrambled photo – it looked like a four-year-old had cut up a photograph of someone's face and stuck it back together back to front and upside down.

feedback with how attractive they each were,* and found that people's looks had almost no impact on whether they had a good time or not. Women enjoyed themselves 75 per cent of the time and men 85 per cent, irrespective of who, if anyone, was better looking. Just to be clear, that means men enjoyed themselves around 85 per cent of the time, whether their date was ludicrously more attractive than them, or ludicrously *less* attractive.

It might be that the people who signed up for the app were people who cared less about looks in the first place, but when photographs were available, the same men and women were much, much more likely to reply if they thought the person messaging them was hotter than they were (and much less likely if the person was less attractive). The very same people were preselecting for something that – once they met that person face to face – didn't seem important.

Of course, having a good time on one date doesn't necessarily mean you are going to have a happy and lifelong relationship together (otherwise I probably wouldn't be writing this book). But it does suggest that filtering based on looks might cut people out unnecessarily.† Plus physical attraction and lust are no guarantee of the sort of companionate love that makes for meaningful relationships in the long term.

---

* Specifically, he looked at the difference of the two people's attractiveness ratings (because how happy you are with someone who scores 'five' on average will depend on whether you generally score 'one' or 'ten').

† Rather than looks, questions such as, 'Do you like scary movies?' and 'Have you ever travelled alone to another country?' are a better predictor of how likely you are to hit it off. In 75 per cent of the long-term couples OkCupid had brought together by 2015, both people answered those questions the same.

What words won't tell you

'I find writing to someone I don't know and trying to be funny absolutely exhausting. It's impossible.' I met Nick in Buenos Aires in a (moderately revolting) hostel. He was halfway through travelling around South America and made me chuckle so much that I asked him for an interview.

'You can't really discern much about someone's personality from the written word, without meeting them. I don't want to write reams about myself or enter into detailed correspondence. I want to meet quickly to see whether we have that chemistry or not. Instead, I meet girls out and about. I have met a couple of girls in cafes. I once went out with a girl from my gym after we got talking in the steam room.

'I tend to schedule first dates on a non-core evening. Non-core evenings are Tuesday and Wednesday. Core evenings are Thursday, Friday, Saturday. Sunday and Monday are not date-friendly – they are for seeing friends or watching quiz shows.

'I don't normally go for a meal – I just arrange drinks so that if it is a total disaster, I can say, "Oh, I'd better head off," after we have each had a drink. Organising to go for a meal on a first date is a bit full on, particularly if you have never met them before – you don't want to find yourself trying to escape after the starter. If it's going well, I can always suggest going for dinner afterwards.

'I don't have a totally textbook first date. I don't go to exactly the same venues each time, but I do have my favourites. One is a dark, candlelit cocktail bar, with shiny and slippery sofas. There's nothing simpler than sliding along one of these sofas to go in for what I call my "cocktail lunge". You can pick up momentum as you slide – dip your shoulder and angle your head in. Easy.

'I usually go for the cocktail lunge at the end of a date. You need to be shameless on first dates and just literally lunge, otherwise the moment is there, but no one wants to induce it.

'Another good thing about this cocktail bar is that it's very close to a train station, so I just walk my date over to the station and off they go. Getting someone into bed is definitely not part of my routine. I'm not that easy. I don't sleep with someone before the fourth date. And I don't always go in for a kiss. There has to be a certain chemistry.'

As I pored over research, I realised that Nick was wise to carefully choose the location for first dates, as where people meet plays a big role in whether they find each other attractive or not. (Remember the wobbly bridge effect from Chapter 1?) For example, you are less likely to fancy someone if you meet them in a hot or crowded room compared with somewhere with fewer people and a more comfortable temperature. Animal research has shown something similar – animals just don't get on as well with each other when they're uncomfortable.

Music is also important. In one (admittedly very small) study, women were asked to read for 15 minutes with either no music, rock music (including The Doors and Jimi Hendrix) or avant-garde music (including John Coltrane and Ornette Coleman), before rating how attractive one of two men were. Women who listened to rock gave a higher rating than women listening to avant-garde music or no music at all. (You know what to play if you bring someone back to yours …)

Music isn't just important when it comes to dating: retailers have long appreciated the power of music and often use it to increase our spending. Music tempo, for example, has been shown to impact how quickly people move around shops, how long they spend in restaurants and how quickly they drink their drinks in a bar. Which reminds me – a bar isn't a bad place to meet if you and your date drink. Research shows that the more alcohol people consume the more likely they are to find someone else attractive. Who knew?!

External factors like temperature, music and alcohol are only part of the picture. We're also influenced by a whole host of biological factors, like how someone smells. Our personal scent is determined by the same set of genes as our immune system, the major histocompatibility complex (MHC), and we generally prefer people who have a different set of MHC genes, presumably because this allows us to create children with a wider set of immune responses.

Pheromones also play a role. Androstadienone is a natural by-product of the hormone testosterone, and to examine its effect researchers sprayed it around half the cubicles in male and female loos.* When people had the choice of any cubicle (because none were occupied), males generally avoided the smelly ones – they sometimes ventured in but would usually walk back out and find an untreated toilet instead. Females, on the other hand, found the scent quite alluring, using the smelly cubicles more than the untreated ones. When psychologists applied androstadienone to the upper lips of women at a speed dating event, they rated men as more attractive compared with women who didn't have any applied. They were also more likely to see them again.

Egg timing

Another factor that plays a significant role, but which I doubt many people think about in the context of dating, is ovulation. Ovulation is when an egg is released from an ovary and travels down the fallopian tube (where a sperm can paddle along and fertilise it). For the few days before as well as during ovulation itself, women are at their most fertile.

Unlike baboons, whose red bottoms swell, the changes in humans aren't hugely noticeable – at least at a conscious

---

* If you're reading this in the US, I'm talking about washrooms.

level. But a woman's fertile window has a subtle yet significant impact on how she looks, smells and behaves. Her features become more symmetrical, her skin softer and her hip-to-waist ratio larger. She is likely to wear more provocative or revealing clothes, consume fewer calories and go to more social events.

Women aren't the only ones who are affected. Psychologists Saul Miller and Jon Maner asked men to sniff a T-shirt that had either been worn by a woman in the few days before ovulation (when she was highly fertile) or much later on in her cycle (when she wasn't). Some were also asked to sniff T-shirts that hadn't been worn by anyone. Miller and Maner sampled and tested the men's saliva before and after, and compared the levels of testosterone, a hormone known for playing a key role in sexual desire.

With one whiff of fertility, men's bodies started to gear them up for desire by raising their testosterone levels – even though they had no idea this was happening. In a similar experiment by the same researchers, men were asked to complete words that could, with the right letter, become sexual or not (e.g. s_x and _ouch). When they sniffed fertility, they came up with more sexual words than those who didn't.

In one of my all-time favourite studies, 18 professional lap dancers in Albuquerque, New Mexico, were asked to record their menstrual periods, work shifts and tip earnings for 60 days (which translates to around 5,300 lap dances). Tips were measured because lap dances don't have formal prices to avoid police charges of illegal 'solicitation' (and the practice of tipping is enthusiastically enforced by bouncers).

Researchers found that, when ovulating, dancers earned about $354 per five-hour shift on average – almost *double* what they earned when they had their period (about $170) and $90 more than during the 'luteal phase' of their cycle (the period between ovulation and menstruation).* None of the dancers had any idea that their menstrual cycle made any difference to their earnings.

As well as making women more attractive (or wealthy), this window of fertility also affects what they go for in a partner. As a woman's body tells her to get on with fertilising the egg her ovary is about to release (or has released), she will often be attracted to men who are taller, more muscular and socially dominant, and have more masculine facial features, voices and bodies.

In short, although there's a huge amount of evidence to suggest that Nick is right – you will only know whether you're really attracted to someone when you meet them in person – it may be that what is causing the chemistry is the enjoyable music, your date's body odour and the fact that one of you is ovulating. When you decide that you would like to see them again, it may simply be your body telling you this would be a good person to reproduce with – which may not be (and often isn't) a reliable indicator of the character of the person, nor how compatible you might be in the long term.

Yet these first impressions, made in less than a blink of an eye and based solely on physical or chemical aspects, are extremely potent and long-lasting. What's more, there's a system in your brain called 'the interpreter'† that likes to

* This doesn't include those who were on the contraceptive pill.

† Neuroscientist Michael Gazzaniga discovered the interpreter when working with people who'd had the two sides of their brain disconnected to prevent epileptic seizures from spreading. He found that the interpreter, located in the left hemisphere, desperately tried to make a consistent story out of behaviours that it couldn't understand because it wasn't connected to the right hemisphere, which was causing them. In one brilliant experiment, Gazzaniga showed the right hemisphere a scary film (if you show something only to the left eye, it will only be seen by the right hemisphere). When the person felt scared, but didn't know why, the interpreter (in the left hemisphere) decided that *Gazzaniga* was scary.

make stories out of all the internal and external information bombarding you to make sense of what's going on. In trying to explain why you have chemistry with one person (and not another), the interpreter (who operates in your subconscious) may not appreciate the influence of room temperature, alcohol and fertility – but instead decide that there was 'just something about that person'.

Pay attention

Meeting someone doesn't just flood you with information about how good they could be to reproduce with. As well as allowing us to get a good sniff, an actual, physical meeting with someone may be key to understanding more about that person and what their values are. Farsideh* explained: 'You can gauge a lot from a person in 3D. You can see how they treat a bartender when they order a drink, how they are when they're on the phone. I don't know if it's healthy to remove that from people. Anyone can give the best version of themselves online.'

Of course, people can, and do, monitor their behaviour over a date or three. But, on the whole, you can also get a sense of how someone treats others when you meet them in person. Going out inevitably leads to awkward moments – the waiter brings the wrong thing, someone cuts in front when you're driving or knocks into you when you're out on a walk. Seeing how someone reacts in these situations can provide a valuable insight into how they regulate their emotions, whether they flare up, shut down, or pause and compose themselves. And that's useful, because, at some point, you're probably going to be on the receiving end of the same reaction.

---

*You met Farsideh in Chapter 4.

You can also get a sense of their attachment style (as well as your own) and whether the two are compatible. People with an anxious attachment style usually devalue themselves, whereas those who are more avoidant often project what they don't like in themselves on to their date.* Unless you pay attention, you won't spot (or stop) insecure attachment behaviours in yourself coming through. Nor will you spot behaviours in your date that might give you a window into how they treat others (and themselves). If you're too busy trying to pretend that you are strong or weak or anything else, you won't be learning about who the other person might actually be. In fact, unless you give the right sort of attention, you won't recognise a potential partner in the first place. And even if you do, without giving your full attention you won't have the conversation you would need to have to explore whether you could have a meaningful relationship with them.

Being open to uncertainty

'There are a number of reasons that have contributed to me not falling in love,' Adam explained, peeling off his fleece. Even though the ship's library was warm, it was discombobulating talking to someone wearing a T-shirt surrounded by icebergs.

'I really haven't exposed myself to the sheer volume of women necessary to find one, probably because most of the things I do are not popular with them. In school, I studied computer science and engineering, where there were all of two girls and they weren't exactly my type.

* They do this to help bolster their inflated view of themselves as strong and independent – which is also how they present themselves to others (at the cost of seeing themselves, and others, accurately).

'The sports I do are not very popular with women, like Brazilian jiu-jitsu and basketball. Having graduated, my job in computer science doesn't lend itself to meeting many women. The normal vehicles where people meet their future partners are closed to me, so it just takes an extra effort.'

Unfortunately for Adam, he was right: a lot of people meet love interests through work or education. In a small UK study, 45 per cent had met while studying and 32 per cent at work.

The numbers were smaller in the US: in the HCMST, 5 per cent had met someone in school, 10 per cent in higher education and about 10 per cent through co-workers.* The percentages were lower for gay men and lesbians, at 0 per cent, 2 per cent and 7 per cent respectively – instead, the vast majority had met online. Meeting through friends was the most popular method of finding love, with 30 per cent of straight couples and 13 per cent of gay couples meeting this way. Not ideal if, like Adam, you want to date a woman, but your fellow students, co-workers and friends are all men.

Spending a lot of time in very male-dominated environments was only part of the problem, at least in Adam's mind. 'I have a lot of things I want to do in life and a lot of interests, so it's easy for me to get caught up doing other things. Another thing I struggle with in terms of meeting women is that in most things I can say, "This is what I want to do, here are the options, and here's how I can get there and do it." With women (and relationships in general), the options are ephemeral. They're ever-changing. So maybe it's partly to do with timing.

'If you're looking at a map, you'll see roads, but the roads don't move – with women, they're flickering in and out. So, it is difficult to predict who I will happen to meet and

* The HCMST is the How Couples Meet and Stay Together study that I talked about earlier in this chapter.

who will happen to be single at a given time. This makes it difficult to know how to meet the right woman. I guess you just have to get out there. It's just not easy to map the steps to do it. There's a large random component in there, which I struggle with.'

Planning for uncertainty

In the late fourth century BC, a Chinese philosopher by the name of Mencius dreamt of starting a new dynasty based on the teachings of Confucius. He took a top ministerial position hoping to have a positive impact, only to realise that he had been tricked by the King of Qi, who then waged war against a neighbouring state and made it seem as though Mencius had advised him to do so. This left Mencius unable to go elsewhere to work for another ruler; his dreams had been destroyed.

Mencius' experience had a huge impact on his philosophy (which, for a thousand years, was read by every schoolboy in China). He came to believe that when we plan, we tend to behave as if there are certain stable factors we can count on in a world that is coherent. We generally assume that the future is predictable – and this impacts our decisions. But Mencius thought the world was *un*predictable and that there were no guarantees of anything.

Instead, he believed we would face victories and heartbreaks that we couldn't have predicted or planned for (or against). Just as close relationships may come to a sudden and inexplicable end, we might have a chance encounter with someone who could change the course of our lives. But whether or not we notice or embrace that encounter may depend on how open we are, whether we build room in our life for all sorts of possibilities and, conversely, how tightly we hold on to our plans.

When I read about Mencius, I was reminded of some of the weird and wonderful ways in which the people

I interviewed had found love – ways that they couldn't possibly have planned or predicted. Henry and Mo met when they were evacuated during the Second World War; Annabel started dating the policeman she met when someone stole her car (and crashed it into a tree); Donny met his future wife on the dancefloor (when he was 16); Ricardo met his girlfriend while canvassing for an election; Marie fell for a shepherd in the Pyrenees mountains; Lynn met lots of hot ladies while protesting; Nick met someone in a steam room; and Sebastian met his wife in a launderette.

Of course, this rarely came up in the research I had been looking at where people were generally given questionnaires replete with boxes to tick ('did you meet: at work, through family', etc.). There were no boxes for 'picking up a policeman during a criminal investigation' or 'chatting up a girl while she was washing her knickers'.* Instead, the more unusual ways of meeting were lost to the boxes marked 'other' or 'none of the above' – in a recent US study of 551 people who were in a relationship, for example, 28 per cent fell into this category. That's more than a quarter who *didn't* meet through friends, work, sport, religion, hobbies, school, family, online, or in a bar or public area. It's a pretty exhaustive list, which leaves me wondering how exactly they did meet …

These random opportunities are obviously difficult to measure, and even harder to plan for (bar optimistically hanging out in the launderette or at crime scenes), but what you can do is try to be open to them when they present themselves. And given that they could come at any time, that means living life as Mencius would suggest – without holding on too tightly to plans and with an attitude of openness.

---

* In defence of the HCMST study, people were asked to provide their 'how they met' stories.

However, this can be deceptively difficult, which is perhaps why a number of people I spoke to commented on the lack of openness in others. Anthony from London* explained: 'I got together with Ayesha, my girlfriend, when we were 19. I remember I was walking up some stairs in the first week of university and I heard this voice say, "Here, didn't you go to the Reading University open day?"

'Then I remembered that when I went to that open day, I spent the whole time chatting to this girl at the back. She went home that same day and told her mum, her dad and her boyfriend at the time, "I met this fella, he was great – he's the sort of person I'd like to make friends with when I go to university." A few months later, we're at the same place (which wasn't Reading). A year later, we were going out.

'You just never know who that person is or where chatting to them will lead. You just never know. I find the possibilities quite exciting. Certainty can be a bit boring. I had other things on my mind: I was taking it vaguely seriously, I was thinking about my future life, I wasn't thinking, "I might bump into a girl who will become my life partner." But I was still open – and being open allows for paths crossing and coincidence.

'Most people, particularly in cities, aren't open to sharing a moment with you. Everyone's moving very fast, everyone has their head in their phone, everyone has headphones on – from a sensory point of view, people shut themselves off from other people. It's a protective mechanism, probably. You're better off closing yourself off because then you can't get hurt, you can't feel silly. You don't make yourself vulnerable to being ignored.

'Ayesha and I were on the bus last week and Ayesha told this random woman that she liked her perfume. She looked at Ayesha as if she'd pulled a dead rabbit out of her bag. But

---

*You met Anthony in Chapter 3.

when you do get them, those moments of interaction with complete strangers are amazing. My favourite is when you see someone dancing in a car, you catch each other's eye and then you both laugh about it. I absolutely love that. I love sharing a smile with someone, or a joke, some banter – it makes me feel good.'

Andy was homeless and often stood outside a supermarket that I passed on my way home from work. I always asked whether I could get him anything and he would request something small, like a sausage roll – 'but only if it's reduced' – or a yoghurt. One day, I gave him a six-pack of yoghurts. He tried to give me some back and when I refused, he decided that he would have two for dinner, two the next day and would give two away to his homeless friends. I was struck by his generosity and asked him for an interview.

'An awful lot of people know they're making the wrong decisions. When I'm on the bus, I see people with their bags on the seat next to them. They stare out of the window hoping not to catch the eye of the person looking for a seat because they don't want anyone to sit close by. They're making 10 different decisions why not to do it. They're thinking, "Don't mind him, he can find another seat, I'll pretend I'm on my phone."

'They spend all their time coming up with reasons as to why to keep their bags on the seat, but if they just made the right decision and moved their bags off the seat, they could relax and feel good about themselves. It's harder to be bad than good.'

Mencius recognised that having an attitude of openness was difficult. When possibility feels more daunting than exciting, the lure of planning and taking control can be irresistible. But Mencius argued that often what you think you can plan and control, you can't. With one exception: you can control whether you try, consistently try, to be a good person or not.

What does he mean by being a good person? In Mencius' mind, this means nourishing your 'heart-mind'* by nurturing both your compassion and reason. Doing this, he believed, would put you in the best possible position to respond to whatever life throws at you.

Trying to be someone who is both good and open doesn't preclude actively looking for love, if that's what you choose to do. But the risk of actively looking without at least trying to develop an attitude of openness is that you may miss out on opportunities that present themselves to you,† or failing to recognise a potential partner even if they did come along.

---

* Mencius used the word 'xin', which means heart, but is home to both our reasons and our emotions. Your 'xin' can contemplate, consider and cogitate, but also feel hatred, joy and love – so it is often translated as both 'heart' and 'mind', or 'heart-mind'.

† And Mencius argues that, the more open you are, the more opportunities present themselves.

CHAPTER SIX

(Not) Looking for Love

So far, I've assumed that everyone wants a romantic relationship, but there's plenty of evidence to suggest that, actually, many people don't.

In a small UK study of single people, 49 per cent of those questioned said they didn't want to find a partner, US research found that 55 per cent of singles had no interest in actively looking and a Japanese study of over 5,000 single people found that more than a quarter said they weren't looking. Being single is now so popular in Japan that the media have dubbed it 'celibacy syndrome'.*

'I'm very happy now,' explained Milena. She was single, Colombian and 45 (but looked 10 years younger). I interviewed her during a work break and we spoke between sipping guanabana juice. 'I've been married, I've had a fiancé, I've had a boyfriend and now I'm alone. I think at this point in my life I'm happier than before, because when you're in a relationship you make sacrifices to make the other person happy. My ex didn't like eating meat, so to make him happy, I let him choose the restaurants we ate in and we never ate meat. He also didn't like eating in bed, so I didn't eat in bed.

'Now that I'm single, I enjoy myself – I eat what I want to eat, I eat in bed, I watch TV whenever I want, I wear the clothes I want to wear, go out when I want to go out. I do

---

* Economists, on the other hand, call it a demographic time bomb. By 2065, nearly 40 per cent of the population will consist of senior citizens because people are choosing not to be in a relationship and are reproducing less and less.

what I want to do whenever I want and whenever it makes me happy.

'It's nice to go out with someone, to the theatre, to the cinema, to have sex. But there are things you have to do in a relationship in order to make it work. You're always stressed about maintaining it, making sure it's always good, always happy. You're always wondering whether they want you, whether you want them, whether they're being faithful, whether it's going to work or not. And when someone has had relationships before, there are more problems – they come with a past, a family, with traumas. When you're single, you don't have any of that.'

Milena wasn't alone in seeing the positives in being single. In the Japanese study I mentioned above, 76 per cent of single women valued the freedom of not being married, while a quarter of single men enjoyed being able to take it easy because they didn't have to be responsible for a family. Half of the people in the UK study I mentioned said they enjoyed being able to choose how to spend their time, while others relished not being 'nagged' by a partner or having to look after someone else. Others appreciated being able to spend money as they liked or had a sense of achievement about coping on their own. Of those over 65, *none* were interested in finding a partner.

'I feel more attractive and better now than when I was 20,' Milena continued. 'I feel prettier and sexier, whereas before I was shy, I was timid. The problem is that the people I meet who are single now have a past. They have baggage. They have their personalities, their aches and pains, their illnesses. By the time you've got to 45, you've lived half your life. It's difficult because they always have kids and their kids will always be more important than your relationship with them. And will you get on with their children? Will their children get on with you? I want to meet someone who does not have much baggage because that baggage will be handed over to me.'

A lot of research into single people compares them with people who are married. Maybe it's a reasonable approach given that, according to the United Nations Department of Economic and Social Affairs, 90 per cent of adults marry at some point in their lives. But, as Milena spoke, she reminded me of an obvious but important point, and a point that is sometimes missed by the 'single versus married' research – single people are a hugely varied bunch. Even within those who are 'single and not interested in a relationship', there is a world of variety. Some hope to be in a relationship eventually. Some don't. Some have a secure attachment style. Some don't. Some are divorced or widowed and grieving the loss of their partner.* Some aren't. Some live alone. Some live with friends. Some live with parents. Some are parents themselves. Some are not parents, but would love to be. Some have strong social support. Some don't. Some are young adults, yet to start work. Some are retired.

It's this diversity that makes it very difficult to generalise. But there is one thing I might be able to generalise about – the single people I interviewed felt a strong external pressure from the 'they' to pair off, as well as an assumption that, surely, they *must* want to be in a relationship. After all, doesn't everyone?

Leftovers

'People often think that what you need to do to be happy is to be in a relationship, get married, have children. And if you don't comply with these rules, then you're not a normal person. Things are changing in Colombia, women are going to university, they're studying, they're more independent, but there's still a chauvinist culture – women

---

* I explore this in Chapter 13.

have to marry, they have to have children. If you have a boyfriend, grandmothers and mothers will ask, "When are you going to get married?" And if you are married, they will ask, "When are you going to have children?" And when you have one child, they ask, "When are you going to have another one?"

'There's societal pressure to get married. But if you break up, who will be the first to criticise? They will. They'll ask, "How could you marry him or her?" The same society that encouraged you to get into that situation will criticise you. Tell me – will society help you pay for your children? Or pay for a psychologist when your marriage falls apart? Does society pay for my rent? My clothes? It doesn't pay for anything. People like to criticise. They like to push you into things, but won't help you get out of them.

'People will always, always, always have things to say about you. I think it's important not to listen. Even if you have the most beautiful children, the best family – people will always have something to say, some comment to make, some criticism.'

Mun Hee from South Korea* lamented how much his mother moaned at him to get married, even though he thought he was very young and wasn't even nearly ready. But it wasn't just his mother who moaned – he explained that, in Korea, marriage is very important and everyone expects you to do it. Meanwhile, Tammy from Chengdu, China,† told me: 'A lot of my friends in their late 20s or early 30s don't have a boyfriend or girlfriend, they are not in a hurry and they are fine. There are more and more people who don't have a boyfriend or girlfriend and they are called *shengnu* or *shengnan*, which means "leftover girls" or "leftover boys". I feel sorry for them because Chinese

---

*You met Mun Hee in Chapter 4.

†You met Tammy in Chapter 4.

culture and Chinese people put them under a lot of pressure.'

It wasn't just family or society at large exerting pressure – plenty of people I spoke to felt pressurised by friends. 'Recently, someone insinuated that I must be a lesbian, that I can't find men attractive,' explained Milena. 'She couldn't believe that I could be happy on my own. I was surprised. I said, "But a relationship, children, don't necessarily give you happiness." She had a husband, children and an unhappy face – and expected me to envy her. But looking at her face, I didn't want what she had.'

Farsideh* who was Persian–American said: 'I have a friend who is always trying to set me up and hear about who I have been dating. The other night, she came out with the fact that she and her boyfriend are so stressed, they haven't had sex in four months. What kills me about it is that it is then pushed on me to date someone, because the assumption is that I must want what they have. Maybe I look at them and think, "Do you know what? Maybe I prefer this. Maybe I would rather have sex with a new person every month for the next 10 years rather than be in a relationship." I'm not saying that is true for me, but that's a decision that I'm allowed to make.

'Any time I vocalise any of these thoughts, people claim that I am some sort of angry feminist who doesn't want to have children. You're not allowed to think these things. There is no thinking outside the box in relationships. Everyone seems to aspire to the same ideal, and you have to find it by the time you're 32. Movies, magazines, culture all feed into that. It stresses me out.'

Khelafi, a French single mother in her 30s, had similar views. 'As a woman there is still the expectation that you need to get married or else you will be a spinster. I think

* You met Farsideh in Chapter 4.

the 1950s expectations are still there even though people act like they are not … It's absolutely preposterous that people think you are supposed to find the love of your life between 26 and 32, and if you don't then you are doomed.'

Some came across an implicit assumption that, if they were still single by a certain age, it wasn't because they had made an active set of choices, but rather that they hadn't been deemed good enough for some reason, they hadn't been chosen. Marie, who lived in the Pyrenees in southern France, explained: 'My parents don't ask whether I'm in a relationship or not. But my friends who have moved to Paris always say the exact same sentence, "I don't understand how a girl like you is single." And I say to them, "I do understand – I want something different. If I didn't want something different, I could already be married with two kids."'

Of course media also plays a role. When sociologists analysed 40 films from a list of the top 200 US grossing romcoms,* they found that single people were depicted as either lonely, miserable, insecure or frustrated.

---

* Their selection process is a bit long-winded, but to give you an idea, they went for films that were released between 1995 and 2005, were set in the present day and had a British Board of Film Classification of 12 years old, Parental Guidance or Universal (meaning suitable for everyone). The films analysed were: *What Women Want, Hitch, Runaway Bride, Bringing Down the House, Sweet Home Alabama, My Best Friend's Wedding, Mr. Deeds, Something's Gotta Give, 50 First Dates, You've Got Mail, How to Lose a Guy in Ten Days, Maid in Manhattan, America's Sweethearts, Two Weeks Notice, Along Came Polly, While You Were Sleeping, Six Days Seven Nights, Nine Months, Down to Earth, She's All That, Bewitched, The Wedding Planner, Clueless, Just Married, Never Been Kissed, Sabrina, Forces of Nature, Serendipity, Kate and Leopold, One Fine Day, Must Love Dogs, About a Boy, 10 Things I Hate About You, Keeping the Faith, Return to Me, Just Friends, The Wedding Date, Picture Perfect, Fools Rush In, and The Prince and Me.*

(Un)happy on my own

The International Social Survey Programme is a colossal collaboration spanning six continents, with over a million people taking part since it began in 1984. In 2012, the project found that 34 per cent of women and 42 per cent of men across 40 countries believed that married people were happier than those who weren't. In Russia and the Philippines, the vast majority (around 70 per cent) believed that to be the case.

Interestingly, women were more likely to disagree with the statement than men (what does that say about the way different genders experience marriage?). In 25 per cent of countries,* around half of the women asked did *not* think married people were happier, whereas this was only the case for men in 5 per cent (namely Denmark and Argentina).† So is it true – are married people happier?

I've come across different findings in different countries at different times, so it's hard to be conclusive, but a lot of research suggests they are. The General Social Survey, a national US survey conducted almost every year since 1972, has collected data from more than 40,000 people. And the married people questioned were happier than people who were divorced, widowed, separated or had never married.

Perhaps that's not because marriage makes you happier, but that it prevents you from becoming less happy as you age. When psychologist Stevie Yap and colleagues analysed data from the British Household Panel Survey (a long-term survey that followed thousands of people every year from 1991 until now), they found that marriage did make people

---

* Namely Slovenia, Portugal, Norway, Iceland, France, Denmark, Ireland, Chile, Belgium and Argentina.

† That doesn't mean the rest of the men and women thought married people *were* happier – some of the remainder were neutral.

happier, but only briefly as people swiftly returned to how they felt pre-marriage. In other words, marriage didn't actually make people happier in the long term. But it did seem to protect them from becoming less happy over time, which is what happened to the single people who took part. It looks a bit like this:

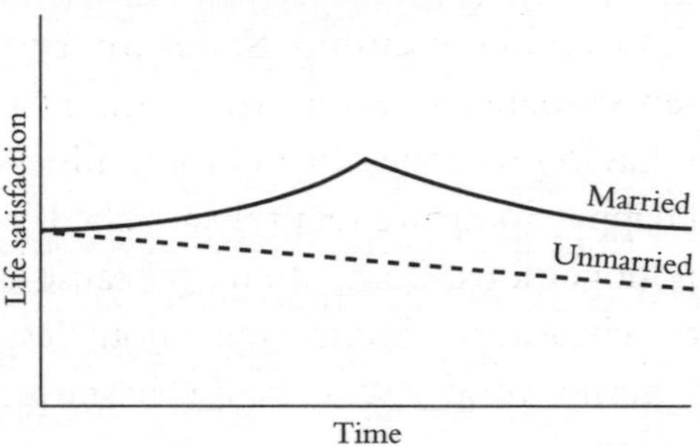

There's not a huge amount of data on couples who are dating or who live together (because most of the research compares people who are single with people who are married). In one of the few studies exploring gay and lesbian couples, those questioned weren't as happy as married couples, but were happier than single people (whether gay or straight). The little research there is on cohabiting couples has found similar results – they're less happy than their married counterparts, but more happy than single people.

If married people are happier (or at least, less unhappy) than single people, it's difficult to know whether that's actually down to marriage or not – some argue that people who get married are happier to start off with. But the British Household Panel Survey I mentioned earlier doesn't support this theory. And if both partners provide emotional support and care, and encourage each other to look after their mental and physical well-being, it feels like a no-brainer that marriage will probably contribute to their happiness (and health).

But not all marriages provide emotional support and care. In fact, some provide the opposite. Just as 'being single' includes an array of people in a variety of circumstances, so too does being 'married' or 'in a relationship'. Some have spent many happy decades together. Some have spent decades living separate lives under the same roof. Some have just had a child (or three) and are exhausted. Some are trying to have a child but failing. Some are trying to figure out what their sexuality is. Some are coming to terms with it. Some are having to cope with chronic illness. Some are healthy and happy. Some are securely attached. Some aren't. Some are monogamous. Some aren't (because one or both partners are cheating). Some are about to end their relationship. Some wonder why they even started. Some are too scared to leave.

As Ricardo* explained: 'I know so many people who are unhappy in their relationships. They don't admit it, but I know them so well that it is obvious to me. Clearly they don't want to say it to themselves or anyone else because they feel they're stuck in that relationship. I've done it myself. I felt exactly the same in my nine-year relationship, but I didn't want to admit it, so I didn't.'

Debating whether marriage makes people happier fails to address something fundamental: the quality of the relationship. In terms of your mental and physical health, evidence suggests that you're better off alone than in a bad one (particularly if there are high levels of conflict). In a study of 8,528 people in China, Japan, Taiwan and South Korea, those who were married but dissatisfied had poorer health than those who had never married.† From upper respiratory tract infections to asthma, cardiovascular

---

*You met Ricardo in Chapter 2.

†This study looked at how people rated their own health, but plenty of studies have reached the same results using objective health measures.

function to peptic ulcers, there is an overwhelming body of evidence to suggest that high-conflict relationships have a significant impact on your mental and physical health. So, it seems getting into a relationship at any cost can end up costing more than you might ever have imagined. But there's some upside – the 'they' will be delighted.

Aloneness

Aside from keeping the 'they' happy, being in a relationship at any cost has another potential benefit – escaping loneliness. Except the evidence suggests that it doesn't always work. In a study of more than 14,000 Germans, 11 per cent said they felt lonely. Although single people were lonelier than those in relationships, couples still felt it, with 6 per cent of women and 1.5–3 per cent of men in relationships admitting to it. (The researchers suggest that the numbers might be higher than this, particularly for men who might not want to admit feeling this way.)

A relationship may or may not help you to escape loneliness, but there's something it won't be able to eradicate: aloneness.

The way you see things is shaped by everything you've experienced up until now. You can try to explain what you're experiencing, and you might be able to give others an idea of what you're going through, but no explanation will ever be complete. No one will ever fully understand what you've been through. This means that we are each alone in our experience of the world, and at times of suffering or in the face of death people report feeling overwhelmingly alone.

'It's an easy mistake to think that you have to fix aloneness by finding someone,' said Farsideh. 'I see people do it all the time. People just want to shut out bad thoughts. They want to pretend it's not happening. "I won't acknowledge that, despite the fact that I am with someone, I feel alone 30 per cent of the time." If you just don't think about it, maybe it's not real.

'A lot of people that are by themselves mistake the feeling of aloneness as something more extreme than everyone feels. Aloneness is a symptom of humanity. It doesn't matter if you're with someone, or not with someone, you will still have days where you think, "Literally, no one understands me. I am scared of dying. I am scared of being alone." There are so many thoughts that go through your mind, they are not going to go away if you meet someone. Why are you so desperate to be with someone? Because you don't want to feel alone in a world where you probably are.'

Psychologist Irvin Yalom believes that aloneness is the driving force behind many relationships and that, in order to avoid it, people often use others, treating them not as a person, but as a tool or piece of equipment.* People usually don't recognise that this is what they're doing. Instead they subconsciously go for someone who helps them avoid aloneness somehow, perhaps by protecting or adoring them, or making them feel powerful. But this creates an entirely different sort of relationship, one in which the other is being used.

It's something that interviewees also commented on. Andy† said: 'Let me ask you a question. What is the opposite of hot? Cold. To truly understand what hot is, you have to understand what the opposite is. Let me ask you another

---

* Yalom is actually talking about what he calls 'existential isolation', but that's a bit turgid so I'm going to adopt Farsideh's language and call it aloneness. The main difference between 'existential isolation'/aloneness and loneliness is that aloneness cannot be avoided; it's a necessary part of our experience as humans. But so is connectedness. As humans we are each entirely unique and entirely related, all at the same time.

† You met Andy in Chapter 5.

question. What is the opposite to love? Using. It's the most terrible thing you can do to another person. It's like putting a dagger in. Using stops you from loving, but a lot of the time, people don't even realise they are doing it.'

Nothing you can do will change the fact that you're alone in the way you experience the world, or that, at some point in your life, you may lose the people you love the most and, ultimately, die on your own. But what you can do, Yalom argues, is acknowledge and confront your aloneness, and then share it in a way that allows for love to compensate for the pain (without using people like pieces of equipment).

Unless you do so, you won't reach out to others, but instead flail at them in an attempt to avoid drowning. This means the relationship will be a matter of survival, not growth, and the person you're with will know that they're not loved, but needed.

Khelafi explained: 'You think that you are broken and you are just waiting for the next relationship to fix it. And it's not fair for the next person and it's not fair to you. You have to be more responsible. The question is: are you really ready to love someone for the rest of your life?'

Psychoanalyst Erich Fromm argues that if you're not able to be on your own, the answer to this question is no.

Who needs love?

There are myriad reasons not to look for love. Being single has plenty of upsides and relationships, even when they're good, have plenty of downsides. In fact, some drawbacks are made worse by being in a good relationship. When osteoarthritis patients had more knee pain during the day, their partners had poorer sleep quality that night – and the effects were strongest among those in closer relationships.

Intimate relationships make us hugely susceptible to the sickness and health, successes and failures of someone else.

Despite the drawbacks, I firmly believe we need at least one, preferably secure, attachment relationship (as well as some social support in the form of friends or family). Although attachments are often romantic, they don't have to be – you can also have an attachment bond with siblings, parents or best friends. Of course, you're likely to share less of your everydayness with a sibling than a romantic partner, but that doesn't have to be the case.

Sally, for example, had been single for over a decade and lived with her sister throughout. She explained: 'I am 53 and single. I don't put any effort into meeting members of the opposite sex. I'm not interested. Some people settle and I think that can be quite lonely. I am happy with my lot, helped by the fact that I have a very close, intimate circle of family and friends. That is enough for me, and I don't have the problems that can come with a romantic relationship.

'Love is many different things to different people in different circumstances. It's almost impossible to describe it because it comes in many forms and many degrees. When people say "love", they tend to refer to love in a partnership or marriage, but actually there are many different aspects to love. We should be very happy if we have any love in our lives, whatever form it takes.'

There are no easy answers. Relationships vary, and even the very best ones have their downsides. But then, so too does being single. Whatever you choose to do, ideally it should be exactly that – a choice. Preferably a conscious, positive and considered choice, rather than one based on fear of intimacy or commitment, or, conversely, of being alone.

CHAPTER SEVEN

Extravagant Expectations

'I think people definitely expect too much from love,' said Chess from New York.* We spoke in the reception of an arts centre, as music reverberated down the corridor.

'Like most things, this concept we've created is impossible. So, people want it, expect it, then get disappointed when they don't get it. Every story since the beginning of time will have had this grand story of what love is in it. Our parents, their own disappointments will be fed into us – and we think we'll right the wrongs of our parents or it will work the other way around, and we think we'll never achieve what our parents have.

'I have had to try and learn – and not in a narcissistic way – that I am a whole person, I don't need someone to complete me. Yeah sure, dating is fun and all that. I've tried. It's failed. The only thing that's going to make me happy is me. I can say that quite easily but actually believing it is really difficult. But that's going to be an easier struggle than trying to find someone, trying to rely on them to be my wholeness and finding that it's a massive disappointment. Again.'

Make me whole

The ancient Greek philosopher Plato conjured up a myth that he tells via one of his characters, Aristophanes. Aristophanes explains that humans used to be round, with four hands, four feet and four ears, and one head with two faces. One day, these peculiar looking humans decided to

---

*You met Chess in Chapter 4.

challenge the power of the gods, leaving the gods in a difficult position.

They considered simply killing the humans, but this would leave them with no one to worship them. So eventually Zeus, the king of the gods, decided that humans would be allowed to live, but he'd cut them in half to reduce their strength. And he did, leaving the poor severed humans desperately searching for their 'other half' to make them whole.* Aristophanes called this – the craving and the quest for someone to 'complete' them – love.

Plato didn't intend Aristophanes' myth to be taken seriously – he was poking fun at romantic love or lust, where one body yearns desperately for another. Yet the phrase 'other half' has lingered and people talk about looking for someone to 'complete them' as if that's actually possible in some way. But no one can complete you. Aristophanes' myth demonstrates the impossibility of it: even if humans were split in two by Zeus, and were somehow able to find their 'other half', they still wouldn't be complete because they could never return to their previous pre-split form. They are, and always will be, two very separate beings.

Yet the idea that there is only one other person out there is pretty pervasive. In a US study in 2013, 31 per cent of men and 26 per cent of women, young and old, married

---

* I should add that these odd-looking humans came in three sexes: male, female and hermaphrodite. That means gay relationships were built into this idea, as they were simply the halves of the original entirely male or female wholes (whereas heterosexuals were androgynes that had been split into two) – not that Aristophanes envisaged exclusive sexual orientations. In Ancient Greece, homosexuality was accepted in the upper classes – although it tended to be one-sided and male. An older, more experienced man could find an adolescent attractive, but the young man could only give in after a decent period of resistance.

and unmarried, believed that each person in the universe has only one true love. In another US study, an extraordinary 88 per cent of 20 to 29-year-old single people thought that somewhere out there was a 'soulmate' waiting for them.

Lust should last

You might agree with these people, or at the very least, think, 'So what? There's no harm in it,' but the fact that some (or many … or even most) think that there's only one person out there for them fills me with horror and dread. That's because this line of thinking often implies a number of other very dangerous and damaging ideas: that love is discovered (rather than built over time); the right person will be perfect (rather than no human being can be); if or when you meet this 'right person', you will 'just know' (so if you have any doubts whatsoever, they can't be right); finding this person is the most important factor (rather than effort or skill);* and the passion will never end. I have problems with each of these ideas, but, for now, I'd like to discuss lust.

Decades of research has found that the older you are and the longer you're in a (presumed) monogamous relationship,† the less sex you're likely to have. Whether or not you've had sex in the past month is not a direct measure of passion – you can feel lust without having sex and you can have sex without feeling lust – but I think it gives us a valuable insight.

When researchers interviewed 6,785 people who were married and living with their spouse,‡ 80–98 per cent

---

* I'll come back to this point shortly …

† To be precise, a lot of the studies examined marriage rather than relationships in general.

‡ As part of the interviews conducted in 1987–88 for the US National Survey of Families and Households.

under 24 had had sex in the past month, compared with 20–50 per cent of people over 75.[*] It's something that also came up in interviews. For example, Noel, an Irishman in his 80s who had been married for more than 50 years, told me: 'Sex comes into it at some stage but doesn't last a lifetime. It can do, but it is not what binds you together eventually.'

There are a whole host of reasons that sex tends to dwindle. Pregnancy, having young children (especially if both partners are working), ageing, diminishing energy and poorer health are usually bad news for passion. Plus, if you start to have less sex as a result of, say, ill health, pregnancy or small children running around and devouring your time and energy, you might get used to it – and then fail to have more if/when your circumstances change.

Then there's something psychologists call 'habituation', which means that the more you encounter something (such as a spouse), the less exciting you find it (or them).[†] A number of studies have explored this (psychologists have even gone as far as monitoring people's genitalia to see what happens when they watch an erotic film on repeat …)

---

[*] The reason for the range in percentages is that some people chose not to answer the question. Do you blame them? It's pretty personal. But some of the people who didn't answer had partners who did, so researchers could still figure out how much sex they were having. So far, so good … but in some cases, neither partner answered – and that's why there's a range – we just don't know what not answering the question means. (Although there's evidence to suggest that the older people didn't answer because they weren't having sex.) But even if we look at the data with all the unknowns, the trend remains the same – the older people were, the less likely they were to have had sex in the past month.

[†] The theory is that, when you're repeatedly subjected to a 'stimulus', your response to it (whether that's excitement, fear or dread) will consistently decrease.

and the findings are relatively consistent. As novelty wears off, so too does desire.*

If it's right, you'll 'just know'

The idea of 'just knowing' cropped up in a number of interviews. Some who were looking for love told me that, if it was true love, they thought they ought to just know. But as I interviewed more and more people, I kept hearing evidence that contradicted this – people either confessed that they *didn't* 'just know', and went on to have successful long-term committed relationships, or admitted they *did*, but went on to discover they had got it very wrong.

I asked Robert for an interview when he came round to fix my boiler. He was very slight with mousey brown hair and an unusually high voice. We spoke over a cup of tea at my dining table. 'I think we've been married 42 years now. We met '69, it was – we were 22. We've been together all this time.

'I guess we're a very close family. We've got two boys, both with partners. Quite a good family life, we have. A lot of it is just give and take. You have to think more about your partner than what you do of yourself. I wouldn't do anything unless it was between both of us. Same as my wife. Everything's together really. I never feel like I need time out.

'I don't think I knew I wanted to marry her when I met her. Back then, you used to court for about three years

* Although I should point out that many of the studies exploring this are flawed. Instead of following people in real-life relationships over many years, they get people into laboratories and show them erotic films: they're not actively engaging with another human being, they're passively watching. So, we might not be justified in extending these results to relationships. But it's the best research we have and it suggests that excitement, like novelty, wears off when we get used to something or someone.

before making any decisions. So, we courted for three years. First it was twice a week, then it was three times a week, then four. We started getting closer and closer. I don't know why I started courting her, really. She's quite attractive, she's got long blonde hair – I think that probably pulled me in. I don't think I ever "just knew".'

Meanwhile, Dani,* who I met in the south of France, 'just knew' before discovering that she'd fallen for a pathological liar. And Sophie, who I sat next to on a flight to Spain, also 'just knew', but with someone who started stalking her. 'With my ex, I knew. We were like *booooom* immediately. After two months, we moved in. We were exactly the same, he was like my soulmate. He was the one I was meant to be with.

'He would get all stalker-ish, constantly ringing my phone a hundred times. Then he put on this spyware app that recorded my phone calls for a week.† It was one thing after another. If my phone rang when we were together and I didn't answer it, he'd say, "Why aren't you answering it?" Sparks would fly because that mistrust was already there. Then I'd pull back.

'When we broke up last, he said, "I'm not worried, we'll still end up together." I don't even understand why I think we were soulmates. But I do think he's the one.'

Whether or not you think you 'just know' can have little to do with whether the relationship lasts in the long term. You might be convinced that you've met 'the one', when the only thing you can really know is that you've encrusted the human being in front of you with glitter (remember the salt mines of Salzburg?). You'll only know whether you can or want to have a meaningful relationship with someone once the sparkle subsides – and that takes time.

---

*You met Dani in Chapter 1.

† This isn't just stalker-ish, this is stalking – which I explore in Chapter 12.

On top of that, when we first meet people, we're often influenced by factors that have nothing to do with whether that person, or a relationship with them, will be healthy or long-lasting.* And what we *think* we feel about someone may actually have very little to do with them. You might be quick to decide that this is the person for you because you have an anxious attachment style and crave closeness, or you might be picky and have doubts as a way of avoiding intimacy and commitment.

The sort of love that makes for a meaningful and long-lasting relationship (in other words, companionate love) requires *real* intimacy and knowledge, which in turn require experience, understanding and time. And that means that the idea of 'just knowing' is dangerous as it can lead to disastrous decisions. As philosopher Simon Blackburn said when I interviewed him in Cambridge, 'Love requires a certain timescale and a certain history. And if the whirlwind doesn't lead to that timescale and history, it wasn't love, it was just a delusion or a fantasy of the moment. And you only find that out with hindsight.'

They kiss and love each other forever

'I think love is when two people meet, and they go on dates, they like each other, they fall in love and they get married. Then they kiss and love each other forever.' Gemma was Greek and had thick, wavy hair down to her knees, which she played with as she spoke. I met her at a birthday party and her mother looked on with pride as I interviewed her. Gemma was 10-years-old.

---

* I looked at how this works in Chapter 5 – an example is that we seem to sniff out people who have immune systems that are different to our own (presumably because this allows us to create children with a wider set of immune responses).

'I have never been in love. But a boy I know has been in love with me. I did not love him because I am too young. I knew he loved me because he told me. He said, "I love you." And I said, "No, you can't love me." I said, "I'm not interested, I'm too young, so no thank you, I don't want you." He said, "Well, I love you, and I can't stop loving you." He sent messages. He gave me flowers sometimes. I was never tempted. I was not interested.

'One day, I would like to fall in love. Maybe. Because I would like to have kids and have a nice family. And I would only like to do that with someone that I was in love with.

'I think love is a good thing, but not if they are cheating on you. If they cheat on you, that might mean they like someone else. If I was in love with someone who cheated on me, I would dump them straight away. I'd say, "Get out, you're not coming back." That's what I would do.'

Gemma was one of the youngest people I spoke to, and I was interested to hear what she expected from love and relationships. At only 10-years-old, she thought love meant falling in love, getting married, kissing and loving each other forever – and that the person she fell in love with wouldn't like anyone else. This incredibly clear idea of love got me thinking – where do romantic expectations come from?

Academics have come up with an array of explanations: capitalism, individualism, consumerism, the decline of religion, advertising, romanticism and growing social isolation (the argument being that we now look to romantic love to provide what an entire village did).

I suspect these explanations are all partially right because, as individuals, we're influenced by lots of different things, from the cultures (and sub-cultures) we live in to our family, friends, colleagues or classmates. On top of that, these factors interact with each other (culture might encourage us to spend more time with work colleagues, say, or friends might introduce us to different sub-cultures) and shift with time.*

What we expect from love is the result of a rich soup of influences – and this makes asking where unrealistic expectations come from an impossible task.† Maybe that's why, when I asked people this question, they zoomed in on one influence – film.

I met Terri in Denver airport. She was sitting nestled in a doorway, waiting for a flight home, when I asked her for an interview. She was a student in her 20s and had decided to go on a short trip on her own to get some space from her boyfriend.‡

'I love movies … and I notice that I will compare my life to them sometimes. I will go to my boyfriend and say, "Why aren't you more romantic with me? Why don't you do this … and this …" and he just has no idea what's coming at him every time. You would think that he would learn by now, poor guy.

'I really want to have a family. I know you can have families without getting married, but I always dreamed of the wedding and that adventure, and for some reason, I still have faith in it, I am not really sure why. I wonder why I do sometimes. My parents are divorced, my mum's parents are divorced, my dad's parents are divorced and it's everywhere. But for some reason I still feel like I might be able to be the one that does it right. I know way more people that are divorced than married and those

---

* I'm leaning on the work of psychologist Urie Bronfenbrenner, who argued that humans were subject to five levels of influences: microsystem, mesosystem, exosystem, macrosystem and chronosystem. But I've tried to keep things a little simpler by avoiding those unwieldly descriptions.

† A bit like asking who, how and why we love …

‡ You met Terri in Chapters 2 and 3.

marriages that do exist, I can only count on one hand how many are like truly happy relationships that I aspire to have some day.

'I was raised very much a feminist by both of my parents, which I think is really good … Other dads would call their little girls "princesses" and I'd ask, "Daddy, am I your princess?" and he'd say, "No." He would tell me, "You don't need a prince to save you – you save yourself." I remember my favourite princesses growing up at the time were the ones who were the least distraught, who had interests, who were intelligent and who fought for what they believed in. So, I think that characters like that are really important.

'I wrote a paper once on how I thought Disney contributed to high divorce rates, and the really high expectations that are set for girls as they are growing up … the fantasy, love and romance. And then on the flip side of it, boys are raised with porn essentially. So these girls go into relationships expecting this Disney fantasy, and the guy is expecting to go in and get a threesome … and they totally don't align. It doesn't work and I think that it really hasn't done our generation any good.'

When sociologists analysed all the G-rated films* that made $100 million or more in 1990–2005 (many of which were made by Disney), they found that (heterosexual) romance was either the main or secondary plot for 75 per cent of them. Only 10 per cent had no reference to romance at all.

Love was often portrayed in these films as transformative of the lovers and the world, as well as being potent and magical. Literally, magical. The lovers were surrounded by flowers and fireworks, music and magic, ballrooms and butterflies, and fell in love in extraordinary circumstances: flying through a starlit sky, riding on a sleigh in a snowy

---

* G-rated means that the film was deemed suitable for a general audience – including children – by the Motion Picture Association of America film rating system.

night, or on discovering how beautiful the earth was (having previously been a mermaid ...).

Friendship, on the other hand, didn't get any of this special treatment. Instead, it was relegated to providing a splash of light humour amid the serious work of falling desperately in love (think Pumba in *The Lion King*). Ironic given that friendship (or companionate love) is the basis of successful long-term romantic relationships.

What's particularly pernicious about this is that these films have a huge reach (at least in contemporary Western society) – only 1 per cent of US mothers surveyed said that their child had *not* seen any of these films,* and half had seen 13 or more. On top of that, children often watch films over and over again – that's falling in love to flowers and fireworks on repeat.

In a separate analysis of 25 Disney films, the vast majority featured couples falling in love at first sight – usually in a matter of minutes.† If they were taking it slow and got to know each other a little more, they might take all of a day to do it. Meanwhile, none of the films featured much, if anything, about how relationships are maintained. Instead,

---

* Based on a survey of more than 600 American mothers of three to six-year-olds. Huge numbers had seen individual films – for example, 85 per cent had seen *Shrek*, and 88 per cent of girls had seen *Beauty and the Beast*.

† Of the 23 that featured romance, 18 had characters falling in love at first sight – that's nearly 80 per cent. This study looked at films released between 1937 and 2000. There was some overlap in the films analysed in this study and the last. The films analysed in this study were: *Snow White and the Seven Dwarfs*, *Pinocchio*, *Dumbo*, *Bambi*, *Cinderella*, *Alice in Wonderland*, *Peter Pan*, *Lady and the Tramp*, *Sleeping Beauty*, *101 Dalmatians*, *The Sword in the Stone*, *The Jungle Book*, *The Aristocats*, *Robin Hood*, *The Fox and the Hound*, *The Little Mermaid*, *The Rescuers Down Under*, *Beauty and the Beast*, *Aladdin*, *The Lion King*, *Pocahontas*, *The Hunchback of Notre Dame*, *Hercules*, *Mulan*, *Tarzan* and *The Emperor's New Groove*.

they featured couples falling in love, getting married and living happily ever after.* The message was clear: meet the right person and everything else will be banquets, ballrooms and butterflies.†

Perhaps because I also write for children, I've always been acutely aware of what we're teaching them through literature and film (and other media). Young people are very absorbent, particularly in relation to things they don't have first-hand experience of themselves (like romantic relationships). On top of that, they're not always encouraged to analyse storylines in the way you might as an adult (not that all adults do this, of course).

So, when they're told, repeatedly, that they will know they're in love within a matter of minutes and that it will be transformative, magical and everlasting (without any evidence of effort), that might just influence the way they think love and relationships work.

If Hollywood is listening …

The people I interviewed didn't just complain about films for children, they were adamant that films for adults were also to blame. I met Carlos in Venice, Italy (but he was actually from Trieste, very close to the border with Slovenia). He was in his 20s and rifling through the navy rucksack on his lap when I asked him for an interview. He explained that he used to be deluded about love and that Hollywood films made him think that 'when you have the

---

* If you're after a Disney film where they don't fall in love at first sight to show to your kids, of the 25 analysed, you have a choice of three: *The Rescuers Down Under*, *Mulan* and *Tarzan*.

† In defence of Disney, some of their more recent films don't have a happy-ever-after or focus as much on romantic love – but many do. And in any case, plenty of kids watch these older films and many adults grew up devouring them, over and over and over again.

right one it is going to be perfect. And you just know that it is just the right one because everything is perfect. But then I guess I grew up and I understood that it just doesn't work that way and I would like Hollywood to stop selling these stories … a lot of people have too many expectations that are just wrong and it is going to become a very big delusion. Or maybe it is part of the process of growth that you come to understand this.'

He paused, scratched his head and told me that, now that he thought about it, he still had friends that didn't understand that and were still struggling to find 'the one'. 'It's just not healthy,' he said, waggling his finger furiously. 'But you have to learn it the hard way. I enjoy a lot of those movies … but it's just creating a lot of delusions. If Hollywood would stop doing that it would be easier for a lot of people … I don't know if Hollywood is listening, hopefully, and in that case – "Just do something else!"'

It would be great if Hollywood was actually listening, as Carlos wasn't the only person to complain. Farsideh* from Iran conjured a delightful image when she said, 'Romantic comedies do not show you someone having to look after their girlfriend when she has diarrhoea, which is real love.'

And Alexandru from Romania† said: 'I used to love watching all these romcoms and sitcoms and stories about romantic engagement. I used to binge watch them in a weird way, I don't know what I was getting out of them to be honest. Maybe as an avoidant it was, "I can just watch from afar, it doesn't affect me at all, I can get all my needs satisfied." But the moment you have been in a couple of adult relationships, that whole way of characterising romance is so false.

'I can't watch these movies now. It does not capture the loved experience of being in a romantic relationship and

*You met Farsideh in Chapter 4.

†You met Alexandru in Chapter 3.

the fantasy that it offers in its place to me is so hollow that I derive no joy from it. I can definitely suspend my disbelief and go and see an action movie, watch one man take down the world and know that it's stupid on some level but enjoy it. But watching some totally hapless couple fall in love, think it is the best thing that has happened to them and walk down the aisle – that to me is total buffoonery and doesn't offer anything helpful about how to think about relationships or how to really relate to other people at all.

'In fact, I think it probably does the opposite and is quite destructive because it makes you think that this is how it should be if they come into my life … Everything is going to be better and they will make me whole. It's actually pathological and dysfunctional, how they talk about love, and it's weird that we insist on this idea of it. It's borderline irresponsible and not to sound too old but this stuff really bothers me.'

Hollywood wasn't the only industry under attack – people also criticised Bollywood for similar reasons. Shortly after I'd finished an interview with a dentist called Ahmed, he yelped and said: 'Wait! I can't believe I didn't talk about Bollywood!' He then went on to explain how Bollywood films had taught him everything he thought he knew about love – until he realised it didn't quite work like that.

I couldn't find any research analysing the romantic content of Bollywood films, but the academics I came across thought that love and romance were as central to Bollywood films as they are to their Hollywood cousins. And, although Bollywood couples usually have to take their extended family into account in a way that those in Hollywood don't, storylines almost always feature a pair of lovers finding fulfilment against all the odds.*

---

* I'm relying on the knowledge of academic Sangita Gopal here, as my knowledge of Bollywood films is minimal.

Pornography

I can't talk about film without at least mentioning the sexually explicit, particularly given that Terri argued boys are raised on it (while girls are raised on Disney). In its 2017 review, one porn site reported that it had 81 million daily visits* – that's 28.5 billion visits a year. The site claims that each of these billions of visits was made by an adult aged 18 or over. I can't say I'm entirely convinced ... But whether or not under-18s are using that particular site or not, there's plenty of evidence to suggest that they are consuming porn. A Swedish study of 718 high school students, for example, found that this was true for 98 per cent of males (and 76 per cent of females).

Amazingly, while most of the young people questioned were adamant that it didn't have any impact on them, most also believed that it *did* influence their friends. Other studies involving interviews with 14 to 23-year-olds suggest that it does have an impact – or at least, young adults believe it does. A theme that came up over and over again was how porn led young people to accept traditional gender roles, specifically women being subservient to dominant men.†

Does it really matter?

The impact of film (or media more generally) is difficult to measure – not least because it's likely to influence different people in different ways. A young person with less relationship experience might be more absorbent than someone who has been married for 50 years. People also

* That's more visits per day than there are people in the UK.

† Another theme that came up was the double standard: women who had multiple partners were seen as promiscuous, whereas men who did the same were revered. Females also talked about feeling inferior to porn stars, while males worried about being able to perform.

watch films differently – some do it irregularly, others compulsively; some watch a variety of genres, others only watch romcoms (perhaps to reaffirm their fantasies of how love should work).

Yet there's enough evidence to convince me that film, or at least media more generally, does play a role in the way people think and how they act, both in and out of love. Research commissioned by the mental health charity Mind found that 37 per cent of men and 15 per cent of women questioned decided to get professional help after seeing a storyline featuring a character with a mental health problem. Although this relies on what people say, there's little question about causation here: in relation to mental health, positive representation in the media played a crucial role in people deciding to take positive action.

So isn't it a shame that, when it comes to relationships, so many storylines do the opposite.

Falling short

Irrespective of how they come about, extravagant expectations can be disastrous. In a small but classic study, psychologists Norman Epstein and Roy Eidelson found that people with unrealistic expectations of romantic love were more likely to want to end their relationship (instead of trying to improve it), had less hope that therapy would work for them and had fewer satisfying relationships.

In another study, people whose (unrealistic) expectations weren't being met were less committed, less invested and less satisfied in their relationship. They also felt that the other options out there were more attractive.

These studies didn't follow couples over the years, so they only provide us with snapshots in time – and this makes it difficult to say whether their expectations caused problems in the relationship, or vice versa. Imagining a perfect alternative is likely to make your current relationship even less satisfying, but equally, a miserable marriage might encourage you to fantasise.

Yet there's evidence that unhelpful expectations can be very robust, and that relationships and life experience might not be enough to change them. In one study, psychologists found that neither age, gender nor number of past relationships had any impact on romantic expectations. In other words, being in a relationship or, at the very least, being exposed to other people having them, may not change your expectations. (I should point out that this study only looked at 18 to 28-year-olds – perhaps 10 years of relationships won't change your thinking, but 30 years will …)

Meanwhile, when more than 500 US students were questioned about their beliefs in romantic love, and then again eight months later, their views were very stable over time. This means that whatever was going on in their life during that period (including break-ups or starting new relationships) did not shift what they thought love ought to be like. (Ideally, we'd have comparisons over a longer period, but sadly the research doesn't exist.)

One more thing – in the Epstein and Eidelson study I mentioned earlier, the beliefs of each partner were unrelated. This might mean that, if you have unrealistic expectations, it won't matter what your partner does, they won't be able to change the way you think. Or, if it's your partner who has them, you may be the one who is powerless. You might have a phenomenal relationship – but if you or your partner expects perfection, what you have won't measure up.

Shake off the dust

Perhaps part of the reason unrealistic expectations tend to stick around is that they're not always conscious, or at least they're often unexamined. You can unknowingly pick them up along the way, like mud on your boots or dust on your rucksack. So, you might think you have completely reasonable and realistic hopes for relationships, when actually, deep down, you don't. You can consciously think human beings are imperfect and relationships take work while secretly believing that when you meet the 'right person' everything will be easy. Or you might, like Alexandru, devour romcoms and buy into the idea that 'love will conquer all', while subconsciously thinking that love can't, in fact, conquer anything.

Most of the research focuses on conscious expectations* – which means we don't know much about the ones that sit below the surface. But Rakel, who I met in Sweden, talked about different layers of expectation brilliantly. She was a student in her early 20s and lived in a place called Upsala. She explained it was the fourth biggest city in Sweden – but quickly qualified this by saying that the total population of Sweden is smaller than the population of London.†

'In Upsala, there are much fewer options, everyone knows everyone and they have lower expectations – people tend to stay in Upsala because that's where they come from, and they're comfortable with where they are and what they have. Most of my friends from home are getting together with people who live two streets away. My brother has just had a baby. He is 22 and lives five minutes away from my parents in a nice house. My other brother still lives at home. My best friend has been dating the same guy since she was 14. I also dated him for a month. I have been to 20 weddings and I am 20-years-old.

'My close friends and I often talk about expectations and the way they change our experiences of things. When I go

* Well, they are easier to measure.

† The population of Sweden is actually larger than London, but only slightly.

to see a film, I tend to think it will be bad because then it usually exceeds my expectations. It's the same with life – I'm cynical because then things can turn out better than I expected. Most of my Swedish friends are the same.

'A lot of people buy into the idea that there is a perfect relationship out there. Deep down they hope that someone will come along and sweep them off their feet. It's probably because we've been indoctrinated that the perfect person will come at the perfect time, it will all be perfect. When life is shit, it's nice to hold on to that fantasy. It can be subconscious. You can be realistic and say, "Maybe it won't happen," but at the same time, subconsciously hope that it will.

'I have several layers of consciousness. I have this unconsciousness where I just feel. Then there's the logical consciousness. I always try and check in with both. Each of them has its own expectations – whether realistic or not. It's hard for some people to be reflective and think about any of this – they don't want to because it's hard, ignorance is bliss and they would much rather go on living their life.

'But if your unconscious believes that you will meet someone perfect who will solve everything, maybe that's a good place to be. I don't think it's an option for me because I'm too realistic. But maybe it can be reassuring because you don't worry as much.'

My fear is that, although subconsciously clinging to unrealistic expectations may in some way provide comfort, it does so at an enormous cost – forcing love, relationships and flawed human beings to labour under impossible expectations.

It's how, not who

In order to try to get to the bottom of the relationship between love, film and literature, I wrote to Jack Zipes, a pre-eminent American academic I'd studied some years earlier when researching children's literature. I attended a lecture he gave when he was next in London and asked why he thought people were attracted to unrealistic portrayals of love. He said: 'Fairy tales and certain films are reflections of what we lack. We lack love. And that, I think, is really key to understanding why people are attracted to these films. It's really difficult to have that wonderfully full, erotic, trusting relationship with someone.'*

Part of the appeal of unrealistic expectations is the hope of a saviour and the absolution of any personal responsibility. All you need is that one 'right' person, your other half, your 'soulmate' and everything will be perfect. But I don't believe it works like that. And I'm in good company because a number of great minds think that love is less about the object of your attention, that one, right person, and more about the process of loving – a process that according to philosophers Nietzsche, Aristotle and Montaigne, takes study and skill.

Psychoanalyst Erich Fromm likens waiting for the 'right person' to wanting to paint, but instead of learning the art, lamenting that you haven't found the right object yet. He distinguishes 'falling in love', which he sees as a very temporary state, to 'standing in love', which he argues is a decision. It's not something you fall into – you actively choose to stand in it. And it's standing in love that's the hard part, the part that takes learning and effort.

Although Fromm provides hints as to what's required to learn the art of love, he deliberately avoids explaining how to do it (a bit like Aristotle refusing to tell us how to be a good person … as much as I would have loved for them to

* He also made another important point – that romantic storylines also often involved women being saved by men – and usually a man on a white horse …

have provided handbooks on these subjects, these thinkers seem to want us to figure it out for ourselves).

What Fromm does say is that love shouldn't be specific to that one special person – instead, it's a way of relating to the world. In fact, he argues that if the only person you love is your partner and you're indifferent to everyone else, what you're experiencing isn't actually love at all.

As I reflected on the evidence and the conversations I'd had, I started to wonder whether part of learning the art of love is to completely recalibrate unrealistic expectations and, instead of seeing love as a matter of finding your 'other half', seeing it as part of a much larger faculty, one that takes learning and skill.

Unrealistic expectations oversimplify relationships. They also provide a point of comparison – except humans are flawed and imperfect, whereas expectations can be as perfect and flawless as your imagination can muster. That means that when you compare the two you will inevitably be disappointed and there may be nothing that your partner can do to change that.

Love is not an all-powerful solution to the problem of finding meaning, security and happiness in life. There is no one person out there for you or anyone else. Relationships are built, not found. They are made up of fallible people, and to expect anything else is to set yourself, and your relationship, up for inevitable and inescapable failure – and, perhaps more importantly, to deny yourself the opportunity to learn one of the most important skills in life: the art of loving.

CHAPTER EIGHT

For Better or Worse

'To me, love is a commitment in itself …' I met Noel, an Irishman in his 80s, walking along the beachfront with his wife, who had severe dementia. They'd been married for over 50 years.

'Having chosen somebody and given your life to them, you go with it and accept the good and bad. Whether that is love or partnership or what it is … Sometimes it is the hard times that tie you together and sometimes the easier things are, the more you are liable to drift apart. For us she accepts an awful lot of what I am … I think that we started off with religion and getting married, for better or worse.'

Throughout the interview, Noel's wife interrupted to ask who I was. Each time, he would pause, turn and patiently answer her, before returning to our conversation. He struck me as the embodiment of 'for better or worse'.

'It is accepting the good and the bad and considering how much worse off it could be … Never taking any disagreements too seriously … realising the other person has feelings and problems and trying to see both sides of any arguments really rather than being too dogmatic about things. It is an acceptance of being different and moving on. Tolerance is an awful lot of it. And think long and hard before you decide to change. Hasty decisions are decisions made after a couple of bottles of wine and not the most suitable for the long term. I think that people don't always consider what the long-term implications are of changing and, to me, sometimes better the devil you know than the devil you don't know. The grass might be greener, but you don't know what it's like in winter …

'We do know people who have split up and why, we don't know. We have been so amazed that it's happened, but it does. Not always because of somebody else coming along, but just deciding "maybe life would be better on my own". But I don't think it would, though I have effectively lost my wife now. That's a bit difficult, but it won't get any better either, so we are here.

'Younger generations underestimate the amount of work that's involved in maintaining a long-term relationship. There is a different attitude to life today and they live very much in the moment and if the moment is not right, then the future is not – instead of saying that moment has passed and the future is often brighter, or it can be. There is too much letting the moment affect a lifetime and deciding to walk away from it … and it is probably easier to do it nowadays.'

Commitment

Psychologists researching love originally thought that when it came to commitment, the main factors were attraction, satisfaction and love. To be honest, that's probably what I thought before I started interviewing people who had been married for half a century.

Of course, it's easier to keep going when a relationship makes you feel good, compared with one that makes you feel bad – but it's overly simplistic to think that love and attraction alone explain commitment. If they're the main drivers, why do people stay together when things aren't going well? And why do some relationships bursting with both come to an end?

The implied assumption is that, if partners love each other and feel happy in their relationship, they're more likely to work at it. And though that's true in some circumstances, it's possible to love someone (or think you love them, or love them to the extent that you are able to) but still avoid commitment.

Surely a key aspect of commitment is how long you persist when you *don't* feel loving or attracted to your partner, when they've really infuriated you and you'd really rather not look at them for a while? This is what Noel and another octogenarian, Henry, said in their interviews – they vowed to stay together 'for better or for worse', and that's what they did. Staying committed only as long as you feel love or attraction would be more like staying together 'for better'.

Theories of commitment

Psychologists have since come up with a variety of theories of commitment, but rather than getting into the nitty-gritty of each one, I want to tell you what they all have in common: they each share a distinction between the sort of commitment that pulls you towards the relationship (which I will call 'personal commitment') and one that stops or discourages you from leaving (which I will call 'practical commitment').

Personal commitment describes the extent to which someone wants to stay in a relationship, and might include things like attraction to their partner, sexual compatibility, shared values and companionate love, as well as attraction to the relationship.

Practical commitment, on the other hand, refers to external limits, pressures or costs of leaving that might encourage people to stay together. This might include friends and family reacting negatively, financial costs (finding new accommodation/dividing possessions) and any religious restrictions. Again, depending on whose theory of commitment you read, practical commitment might also take into account 'investments', which would be lost or lose value. This doesn't just mean money, but also non-financial investments like mutual friends and the intimacy that partners build and share.

Practical commitment might not sound particularly romantic or exciting, but it's often this that keeps people in a relationship when, in that moment, it might not be a particularly fun place to be. It can also help create a sense of safety and security.

Depending on whose theory of commitment you read, personal or practical commitment might also take into account the alternatives to being in the relationship. This doesn't just mean other potential partners, but other options, like being single. The less attractive these options, the stronger personal/practical commitment is likely to be.

Personal and practical commitment aren't necessarily distinct – and, in fact, can influence each other. People with plenty of personal commitment are more likely to set up practical commitments without necessarily noticing their impact at the time. After all, why would you be conscious of what's stopping you from leaving when you can't imagine going in the first place?

Different combinations of personal and practical commitment lead to very different types of overall commitment. The person who is with their partner because they want to stay in a relationship will have a very different commitment to the person who is sticking around because there's too much standing in the way of them leaving, and different still to the person who feels a combination of both.

And, of course, commitment changes over time. You might start off with personal commitment, go through a bad patch when you rely on practical commitment, and come out the other end to find personal commitment again. The risk is that people who postpone or avoid making practical commitments can find themselves having to rely on personal commitment (which isn't always a great strategy when times are tough).

This was something that both Noel and Henry raised in their interviews – namely, that younger generations might be inclined to end their relationship when things aren't going well, even though, had they persisted, they would probably have gone back to having a happier relationship and the corresponding personal commitment. As Henry put it, '"For better or worse" is what we said – and that's how we've kept it. People take the easy way out in this day and age, they say, "Oh we're not suited any more and I fancy this other lady now – ta-ta."'

Belief

But there's another factor beyond personal and practical commitment. What struck me as I spoke to people of different ages and cultures was the many and varied perspectives of commitment they had and, ultimately, how this impacted their actions.

Some who had chosen not to make practical commitments, such as getting married, believed that the personal commitment they'd made was for life. Barbara,* for example, who lived with her partner but wasn't married and didn't plan to be, said: 'John and I have been together for 13 years and there's no doubt that we'll be together forever. I couldn't imagine anything else. We've been together since our teens. We've grown up together, we started our careers together – that's my life, that's the way it is.'

Others, who had opted for practical commitments that were theoretically lifelong, weren't so sure. Robin (who had just got married having proposed to his partner during

*You met Barbara in Chapter 4. We met by the seaside in the south east of England.

a thunderstorm in Cuba) confided that he didn't think he could 'confidently say that we will be together for the rest of our lives'. He went on to explain that things could change and that he was a realist.

I couldn't help but wonder how their very different views of the commitments they'd made would impact their actions – would the belief that you were committing for life make you more likely to act in ways that meant you did?

The extent to which our beliefs play a role in how we behave in and treat our commitments is what philosophers call the 'evaluative' element in commitment. Some philosophers, like Roger Trigg, argue that our beliefs *do* impact our actions when it comes to commitment. I think they're right.

What influences our belief in commitment?

Avoidant people avoid commitment because somewhere along the way, they have come to subconsciously believe that commitment is dangerous and results in loss. This means two things: they can avoid making positive commitment decisions, like moving in or getting married, but they might also do things that undermine commitment, like becoming interested in someone else. And given what we know of attachment theory, that's not surprising. Avoidant people idealise independence, and committing, whether personal or practical, puts them at risk of being dependent, which their attachment systems don't like.

It's not just highly avoidant people who struggle with commitment. People with a very anxious attachment style also find it difficult, but in a different way. Anxious people tend to fall in love more quickly than those who are secure or avoidant, often before they know their partners well, and this means they're more vulnerable to leaping into relationships with people who disappoint or hurt them. Anxious people want committed relationships but often choose partners (namely, those who are more avoidant) who don't share their desire. As a result, anxious people often believe that they're more willing than others to commit.

People whose parents divorced when they were children can also struggle with commitment and are more likely to themselves divorce. While this may be the result of the impact of divorce on their attachment style, some psychologists think it's because they develop a negative view of marriage, or because they're more inclined to see divorce as a way to end a problematic relationship.* In other words, because they've seen it end, they view marriage as much less of a practical commitment than those whose parents never broke up. The upshot is that they are less likely to be committed and devote time and energy to a relationship – which reduces relationship satisfaction and, as a result, personal commitment.

As well as factors that influence us on a micro scale, global trends may be changing individual beliefs in commitment. In almost all of the 34 OECD† countries (which include the UK, US, Australia, Japan, Korea and a host of European countries), marriage rates have fallen over a 40-year period (comparing 2010 with 1970). In some countries, the rate has almost halved. In the US, for example, 68 per cent of all 20-somethings were married in 1960 – but in 2008, only 26 per cent were.‡

That's only part of the story. Marriage rates aren't just decreasing – divorce is also increasing. With rates as high

* I explore this more in Chapter 13.

† By OECD, I mean the Organisation for Economic Co-operation and Development.

‡ These stats only measure heterosexual marriage. That's because gay marriage was only legalised for the first time in 2001 in the Netherlands, with a number of other countries following suit, such as Canada in 2004, the UK in 2014 and the US in 2015.

as 53 per cent,* it's no surprise that some may have lost their faith in marriage as a lifelong commitment, or the possibility of lifelong commitment in any form. And the more acceptable divorce becomes, the less likely people are to disapprove of it – making it even easier for people to walk away.

Believing in commitment

But, despite all of this, the evidence suggests that marriage is still something that many crave – even if they're postponing or avoiding it. In a 2014 US study, most unmarried millennials (those born between the early 1980s and the 2000s) still said that they would like to make the practical commitment of getting married. And perhaps that's because committing is incredibly powerful in the context of relationships.

Commitment doesn't just stop or discourage you from leaving when times are tough – research suggests that, in the face of threats, it can even bolster a relationship. (Unhelpfully, much of this research has just measured 'commitment' without specifying what type.)

Studies have found that committed people either ignore their partner's bad behaviour (like snapping when they've had a bad day), or see it as much less of a big deal compared with outsiders. On top of this, they're more likely to react in constructive ways. They're also more likely to ignore people they find attractive, as well as devalue them. In fact, meeting attractive people made committed people more tolerant of their partner's bad behaviours – in the face of threat, committed people strengthen their relationship.

* This is the amount of couples in England and Wales in 1976 who married before 20, and went on to divorce within 30 years.

The dark side of commitment

So far, I've explored commitment without considering whether a relationship is worth getting into or maintaining in the first place. Of course, finding reasons to walk away might be a subconscious strategy to avoid commitment – but sometimes relationships are toxic, and personal and practical commitment can only increase toxicity by making people more tolerant and keeping them in relationships longer than they need to be.

So, commitment for commitment's sake isn't the aim. In cultures where practical commitment is high because, say, divorce is unacceptable, couples can simply end up living separate emotional lives in the home while performing joint marital duties outside it. But that might be a relatively positive outcome compared with others. Psychologists John Cuber and Peggy Harroff explain that while a 'stable' pair might mean a fulfilled couple, it can also mean two people who are entrapped, embittered and resentful, living in an atmosphere of hatred and despair.

Commitment can cause people to stick around in relationships that aren't just embittered but abusive, encouraging them to forgive their partner when they harm them. Although this helps keep the relationship going, it does so at the cost of their emotional and physical well-being. What's worse is that, in some circumstances, forgiveness can cause more of the same bad behaviour.*

Don't commit – it's too risky

Given the risks of making a bad practical commitment, philosopher Dan Moller proposes that we don't commit in the first place. In what he calls the 'Bachelor's Argument', he proposes that most of us think of marriage without

* I talk about forgiveness more in Chapter 9.

mutual love with something like horror – yet the love that might exist at the start of marriage can come to an end. So, when we marry, he argues, we're putting ourselves at risk of ending up in a loveless marriage.* And this is a mistake.

In reality, Moller thinks most people won't take his argument seriously because we're rarely willing to accept that we might change in ways that we may not want to believe are possible ('No, I definitely won't get bored in about 10 years' time and fancy the next door neighbour') and because we tend to leave philosophy behind when it comes to confronting daily life. But he also concludes that maybe, just maybe, the benefits of commitment outweigh the risks.

I've come to the conclusion that the benefits do outweigh the risks (and in the process of writing this book decided to get married). I don't think the answer is not to make practical commitments. Rather than avoid them altogether, I think the key is to think very carefully about who, and what, you decide to commit to – whether personally or practically. But to *really* make an informed decision, you have to be armed with information about yourself – and the other person – that may take a bit of time and exploration to obtain.

If and when you do decide to start making practical commitments, you're still faced with the daily decision of whether to try to maintain personal commitment or not (something that will be harder for those with an avoidant attachment style). After all, despite Moller's argument, we usually have some control over whether we end up in a loveless marriage or not. This was brilliantly explained

---

* While I agree that romantic love can dwindle, that doesn't necessarily lead to a loveless marriage. In some circumstances, it can lead to something much deeper and long-lasting – companionate love.

to me by a lady I met in Tuscany, Italy, called Liv. She was in her 60s, had immaculate posture, and spoke slowly and deliberately.

'Love is a decision, not a feeling … I don't think the decision is, "I'm going to fall in love with that man," as attraction and falling in love are animal instincts. The decision comes after that, once you have fallen in love. You can stop it if it's not right, for example, or you can choose to commit for better or for worse.

'If you choose to commit, it is a decision to love your husband when he's disappointing you. We have to love our partners all the way through the relationship – or at least appear to love them. We might not love them on the inside, but if we love them on the outside, the decision is starting. Then it goes through our whole body because what we say and what we think is what we are. It is a decision to make the best of life, to try and have good thoughts about things, because they become reality.

'I don't think culturally we are taught that. Instead we are taught, "Change your mind if you don't like them any more." So we don't stick at it. We don't try if it goes wrong, we always think that we can do better.'

CHAPTER NINE

I Can Laugh Now

'He was having not one, or two, or three, but four relationships with other people. So that was quite unusual. A multi-infidelity.' Marie stopped to laugh, a deep belly laugh. 'I can laugh now, but at the time it was really painful. And I found out because I noticed some signs … some weird behaviours. I knew him, so I noticed something was different, something was changing. He started to have excuses – "I don't feel like going, but you go." I had to do a lot of travelling and he would say, "But you go, don't stop – you should go with your friends." It started to be the norm. It seemed like he didn't want to be with me any more.'

Marie and I sat overlooking waves rhythmically splashing against the algae-coated sea wall, listening to the heave-ho of the moored boats as they rose and sank. 'It's easy to see what's on mobile phones if they don't have a code. So, I went on his phone and then I saw just how wrong it was. He wouldn't want me to get on the computer in case I saw something. And I did. I found some photographs. They were quite explicit.

'First, I tried not to say anything, but I had to confront him. Then the denial started, and he never admitted that he had cheated. Can you imagine? Even the photos … "Oh this was just playing games or making fun or blah blah." But the photos were really quite explicit. Eventually he moved out.

'To this day, he denies there was anything going on. But there was. The types of messages I read meant something definitely was. It's frustrating he didn't have the courage to admit it or apologise.'

She explained (half laughing) that she only knew of four. But there may have been more. 'I was horrified to see that it wasn't just one person … it was like a bad soap opera. And for seven years! I talked with friends and they were all as shocked as I was. They wondered how he could do that, how he had the time! It felt horrible when I found out – it was a shock. I knew something was up but at the same time, I wasn't expecting it. It was frustrating. Disappointing.'

This time, instead of laughing, Marie started crying. She apologised, explaining that she hadn't spoken about this in eight years. 'It's one of life's mysteries to me. Maybe he needed it. Maybe it's just his personality … I don't know how to explain it. Other than that, he was a good person with a good heart. But in the end, it wasn't so …

'The relationship probably wasn't working because he had the need to find other women. I was working a lot. That provides a lot of free time and opportunities. But a normal person would have perhaps tried to break up rather than cheat with four other people …'

She wiped her eyes and tried to regain her composure. 'If I hadn't said anything, we would still have been together today.'

She trailed off into half laughter, half tears and I found myself thinking about what it was that drove him to cheat so extensively. Was Marie right in thinking that it was in part down to his personality? Was it really fair to say that she was partially to blame? Does infidelity take place in happy relationships? What factors (if any) make someone more likely to be unfaithful? Perhaps because I'd felt so upset for Marie, I was determined to find some evidence to show that she wasn't (at least entirely) to blame. And I did. But I also came to appreciate just how complicated a subject infidelity is – and how much of the time, though they may not want to take responsibility, both parties play a part.

There are a number of factors that can make infidelity more likely – including the relationship itself, the personality and upbringing of both partners, and their environment (e.g. friends, parents, work, culture). But these factors don't impact infidelity in a straightforward way. Infidelity involves a series of decisions and each factor impacts each decision in different ways.

First of all, there's the slippery slope before anyone does any cheating (this might include phone calls, meetings or fantasies). Then there's the moment that someone crosses the line; the maintenance of the affair (if it's not a one-off); its disclosure or discovery; and each partner's response. As well as each of these decisions, there are a host of predisposing factors that, according to research, make someone more or less likely to cheat.

Some factors (e.g. relationship difficulties) can make infidelity more likely at each phase. But others (e.g. self-esteem) have a different impact at different times. Low self-esteem might make you more likely to start an affair, but once the affair is taking place, you may find that your self-esteem improves thanks to all the fresh attention – and this increase in self-esteem might make you more likely to maintain it.

Most research focuses on predisposing factors – probably because they're the easiest to measure. Insights into the other stages generally consist of comments or theories from practising psychologists (rather than evidence-based studies conducted by academics). The research also focuses on heterosexual relationships – and, in particular, marriages. But a 10-year marriage with children is a very different relationship to a three-month relationship in your teens – and on the whole, marriage seems to lower the likelihood of infidelity (even compared with couples who live together).

Predisposed to cheat?

In Chapter 1, I looked at studies that suggest men are more likely to cheat than women (it's one of the most consistent findings across decades of research), but it doesn't stop there. Men express more desire to cheat, more willingness to do it and are more active in seeking someone to have an affair with. This means, statistically speaking, Marie was more likely to be cheated on, simply because she was dating a man.

Another way in which people differ from each other is their personality and their desire or need for novel, varied and intense experiences. Psychologists call those who relish new experiences and need constant stimulation 'sensation seekers'. High sensation seekers are more likely to smoke, drink and take drugs, and are more attracted to high-stress careers (such as law enforcement, firefighting or working in the emergency room). And they're more likely to be men. There's a variety of biochemical and neuroscientific evidence to suggest that high sensation seekers don't have a choice about this – they process things differently in their brain.

Unsurprisingly, researchers have found that sensation seekers are more likely to be unfaithful (although to be clear, the studies did not measure actual infidelity, only imagined – so people were asked to imagine circumstances in which they might cheat, rather than whether they had actually cheated). I should add that, while high sensation seekers may have no choice about how their brain processes new sensations, this doesn't mean they have no choice over whether to cheat or not. It only means that the new sensations may be more appealing to them in the first place. Sensation seeking peaks in the late teens and early 20s and gradually declines with age (along with levels of testosterone, which has been linked to a higher rate of infidelity) – so this should be less and less of a factor as sensation seekers age.

Sensation seeking is only part of the picture. There's also evidence to suggest that people are more likely to cheat if they're alcoholic, narcissistic, psychopathic or have a higher sex drive. And there are, of course, personality traits that mean people are less likely to cheat, for example, being 'conscientious' (someone who tends to persevere at tasks and follow rules) or being 'agreeable' (someone who tends to avoid conflict with others and seek harmonious relationships).

'Sensation seeking', 'conscientiousness' and 'agreeableness' are all personality traits – in other words, aspects of personality that some psychologists believe are stable over time. Personality trait theory developed when a number of psychologists began looking into personality and, through their independent research, came up with very similar personality traits. But some psychologists argue that trait theory is over-simplistic and that traits aren't actually very good at predicting how people will behave. Instead, they see circumstances as more important, or at least the combination of circumstance and traits.

So, it's perfectly possible for someone who is a sensation seeker in most areas of their lives to have never cheated and never do so. This may be because they're in a happy relationship, there aren't many other potential partners to tempt them, or because their parents had a long, happy and faithful relationship, which provided them with a stable and consistent monogamous model to work from.

Like many academic theories, personality trait theory is not without its flaws – and being a sensation seeker does not mean you will necessarily seek the sensation of sleeping with someone outside your monogamous relationship. But it might increase your chances of doing so.

The relationship itself

Marie's view that 'the relationship probably wasn't working because he had the need to find other women' reflects a commonly held assumption – that affairs only happen when there's something wrong in the relationship. This is certainly the way Ethan, a paediatrician from Wales, saw it.*

'I have cheated on a partner. Two partners, actually. I didn't have the gall to tell them that I was upset or frustrated or angry with them, so I reacted in much the same way that my girlfriend reacted when I was younger.

'She wanted to do something explosive to scream for help, so she cheated. It was a very tumultuous relationship and we were very unhappy for a long time. It felt in many ways inevitable. Sometimes I think people want to be caught, and sometimes with a level of grandiosity, for example, sleeping with your best friend.'

But studies suggest that, while in some instances infidelity happens where there's something wrong in the relationship, this isn't always the case.

When psychologists Graham Spanier and Randie Margolis spoke to people who were divorced or separated, 70 per cent of cheaters blamed their infidelity on marital problems. But that's only half the story, as the non-cheating partners usually thought infidelity was a cause, rather than a consequence, of their relationship problems. As with anything that relies on people's personal accounts, it can be difficult to know what the reality is. The non-cheating partners may have been denying that problems existed or avoiding them – but equally, the cheaters may have wanted to find a reason to justify their infidelity.

This might be down to 'cognitive dissonance', a theory developed by psychologist Leon Festinger, which suggests that people want their attitudes and behaviours to be consistent. In fact, they like consistency so much that when there's *in*consistency, they often change their attitudes to make them fit with their behaviour. In the case of infidelity,

---

*You met Ethan in Chapter 1.

cognitive dissonance could motivate cheaters to change their behaviour and end the affair. But ... it could also have the opposite effect, encouraging them to believe the marriage was already over, see their spouse or marriage more negatively, or suddenly develop a more liberal attitude about infidelity.

That said, people don't always blame their indiscretions on relationship problems. Psychologist Shirley Glass found that a hefty 56 per cent of men and 34 per cent of women who had cheated rated their marriage as either 'happy' or 'very happy'. In another study, people were asked to think back and remember any marital issues that preceded infidelity. Only 36 per cent said they had relationship problems before cheating – leaving 64 per cent who didn't.*

In short, it's difficult to generalise as to whether relationship satisfaction plays a role in infidelity or not. But what I can say is that infidelity doesn't just happen in unhappy relationships – and whether or not relationship problems are the cause of infidelity, they're likely to be a consequence.

As Marie and I both sat listening to the heave-ho of the boats, a handful of tears travelled down Marie's cheeks, creating silvery lines down her pretty face. I could feel tears building up myself. She wiped them away and apologised, half laughing.

'I don't have any resentment. But the initial shock made me question how I could have been such a fool for so long. We had been married for 14 years and this had been going on for seven. I had been really blind.

'I've been very careful since. It's very difficult for me to trust now. I've had some relationships but nothing serious

---

* In an ideal world, the researchers would have asked people at the time (i.e. *before* they cheated), as this would have avoided relying on their memory. But they didn't.

or important because I'm afraid. It was probably the biggest shock I have ever experienced. I felt really bad … and I don't want to have to go through it again. Maybe someone will blow me away and I will totally lose my defences …'

I suggested that other people may not always be able to eradicate our defences – that we might have to remove them ourselves. 'Yes …' she said, 'But so far, I haven't done that. They are still here.' She laughed before noticing that her friend had arrived and it was time to end our conversation. She asked if I had any more questions – and I asked what I almost always ask at the end of an interview. I said: 'What's the most important thing you've learnt about love?'

She stared out towards the water and smiled. 'Expect the unexpected.'

Cheating is never actually fun

'I'm not really an expert on love.' I met Jessie at a burrito joint in Hong Kong. She was in her late 20s and wore towering patent stilettos with a matching handbag. 'I think the love thing changes over time. I met Roger when we were in our teens, so I was selfish to begin with. I expected things to be the way I wanted them to be.' She paused and looked at me as if debating whether to carry on. 'I slept with other men when we were first going out, but I still loved him.

'Getting married was a deadline for me: I didn't want to be unfaithful when we got married (although I hadn't been unfaithful for quite a while before that). Having spent years being really scared that I would spend my whole life struggling to be faithful, I just lost all interest in it. Planning the wedding with Roger was really nice and I thought, "Why am I doing this to you?"

'The cheating thing is never actually fun. You always think it will be, but the build-up is the best part. It's a lovely feeling when somebody fancies you and wants to sleep with you because you feel hot and sexy. But the actual cheating and having sex part is usually really disappointing because how many men are really great when you sleep with them for the first time? None. Then you feel like a whore afterwards when you have to get dressed in front of them. On top of that, you have the guilt and the lying. Why go through all of that when all you really want is that little frisson that makes you feel hot and sexy?'

Listening to Jessie was a welcome antidote to theories I'd come across which suggested that if you had cheated in the past, you were more likely to cheat in the future. While part of me saw the reasoning behind that, it also left me feeling uncomfortable about how little room this left for personal growth and choice.

What also struck me about Jessie was her ability to change something that she had previously found so alluring into something that didn't interest her. The temptation stayed the same – it was Jessie who had changed. It reminded me that the power is not in the stimulus, but in how we evaluate it.

So how can you change the way you evaluate things? And in particular, formidable temptations, like someone you find attractive?

In a famous series of experiments by psychologist Walter Mischel, young children were offered the option of one marshmallow immediately or two if they were able to wait. Mischel then followed these children, monitoring their progress through life in order to see how their early self-control impacted other areas of their life. The repercussions were enormous. Children who were better at delaying the reward were better at both maintaining close relationships and resisting temptation. But it didn't end there – they thought ahead and planned more, were less side-tracked by setbacks, more self-reliant, reached higher levels of education, had a significantly lower body mass index, were more resilient, better at concentrating and less likely to use risky drugs.

Obviously, in an ideal world, everyone would learn these skills as a small child and then reap the rewards for the rest of their life – but they don't. Children who grow up unsure of whether they will actually get two marshmallows if they wait, perhaps because they can't trust their parents or caregivers, learn that it's best just to take what's in front of them. They don't develop their mental 'brakes', and instead learn to give in to the desires of their mental 'engine'. But there is hope for people who haven't learnt these skills at an early stage. Mischel found that when children were taught simple strategies to help them resist temptation, they fared much better in the face of desire.

One strategy is to avoid focusing on how delicious the source of the temptation might be – whether that's a marshmallow or an extra-marital affair. This is something Jessie had learnt to do – instead of focusing on the sight, smell and touch of a handsome man, she looked at infidelity as an unenjoyable or even disgusting experience. She did this by thinking that the sex would be terrible, the experience would make her 'feel like a whore' and then she'd be burdened with guilt and lying. Temptation suddenly becomes much less alluring when you look at it like that.

Another strategy is to focus on what you get if you don't give in to temptation, whether that's getting two marshmallows or staying in an honest and monogamous relationship. There's evidence to suggest this strategy actually works: when children were shown pictures of what they'd get if they waited, they were able to delay eating the marshmallow by an average of 18 minutes. To be clear, that's 18 minutes of a child sitting in front of a marshmallow but refusing to eat it. I'm not sure I could last that long.*

It's also helpful to focus on the long-term costs of giving in. When smokers were asked to focus on the long-term

---

* Actually, I could. I don't have much of a sweet tooth (with the exception of salted caramel ice cream …).

negative effects of smoking, their cravings dramatically reduced. This is what Jessie did when she thought about the years of lies and guilt instead of the temptation in front of her.*

In the context of infidelity, guilt and lies are only some of the long-term downsides. Others include a higher likelihood of divorce, long-term psychological consequences for any children you might have,† and symptoms similar to post-traumatic stress disorder (PTSD)‡ for your partner. If you can imagine the future vividly like this, you'll be fighting the heat of desire with flaming hot consequences, instead of the cold rationality of the mental brakes – and this may make it easier to resist temptation.

Jessie paused to eat the last of her burrito and I did the same, before asking whether she'd told her husband about her indiscretions. 'I thought about telling Roger at the beginning. Whenever I watched romantic movies, I felt like I had to tell him. But actually it would have achieved nothing because I figured out that it was a mistake and I stopped. He would have been very upset and hurt, and if I had explained it to him the way I explained it to myself in my own head, he might not have understood it. Now it was so long ago, there is absolutely no point.

'Part of me thinks that he knows that I have cheated. Obviously he won't know everything and every time, but

---

* Another idea is to think of the 'scar' it might leave – as Claire (who for years only dated married men) described in Chapter 1.

† I explore the long-term consequences of infidelity later in this chapter.

‡ PTSD is an anxiety disorder caused by exposure to a traumatic, highly stressful or exceptionally threatening event or situation. Someone with PTSD usually has recurrent and intrusive nightmares and recollections of the event(s), and can feel isolated, irritable or guilty. They can also have problems sleeping and find concentrating difficult. Symptoms are generally severe and tenacious enough to have a significant impact on their daily life.

he probably had a suspicion from me going out with girlfriends, coming home really late and hiding my phone so that he couldn't see the text messages that I was getting. Maybe there is just stuff that doesn't need to be said.'

I didn't have the willpower to say no

I met Mifune, from Japan, in a cable car in Switzerland. Having fallen in love once at the age of 22 ('We weren't actually that compatible, but I didn't have the maturity to realise it'), she moved to Indonesia and fell in love again, this time with one of her best friends. Who was married. Neither admitted that they were in love. In fact, they ignored it – until someone, irritated by how much they argued, said, 'Just admit you're in love …'

'Two things stopped me from having an affair with him. The first was that sex was still sacred as it wasn't something I had done with many people. The second was the damage from the previous relationship – the shock that love can end, love doesn't solve it all. So, when he was saying, "We're in love, we should do this, I'll leave my wife," I kept saying, "No, because love can stop. We can't mess up your marriage for something that we can't be sure will last."

'Looking back, I still think it was the right decision. But I'm really surprised that I didn't give in to his passion, particularly as I was desperate to find somebody. I wanted some solidity in my life – and I saw love as providing that.

'A few years later, in Sudan, I experienced a much shorter version of what happened in Indonesia. It's no coincidence that both happened when I was far away from home and my roots. He was half European, half Kenyan. And engaged.'

When Mifune said this, I remembered something I'd come across when researching self-control – namely, that one of the most crippling influences on our ability to resist temptation is stress (which can be caused by loneliness). The brain is quite literally rewired under stress – the amygdala, the brain's smoke detector,* grows in size while our ability to make rational decisions wanes.

'I was far away from home, and we got on as if we had been friends our whole lives. We had grown up in similar places, so we shared that whole vocabulary that you learn as teenagers. His words brought my home to me. There were warning signs quite early on. He would say things like, "If I wasn't engaged to someone, we would be dating," or "You remind me of one of my ex-girlfriends."'

The cable car clonked up the mountain and Mifune took a moment to appreciate the snow-topped peaks. She told me that one weekend she went on a boat trip with close friends and felt relaxed, carefree and loved. It was the antithesis of living in Sudan, where she felt pressured and isolated. When she got back, she went straight back to work and suddenly felt incredibly lonely.

'I remember ringing him and saying, "Oh God, I feel awful." He said, "Don't stay on your own in a hotel, come and stay in my house – I have a spare room here, there's no need for you to be alone."

'I went to his house and it became clear that there was an attraction there. He put a mattress down on the floor so that we could watch a film together and we ended up kissing. I said, "Look, we can't do anything" – but this time I had much less willpower. We fell asleep and at some point during my sleep my mind gave up and decided that I had already gone too far. We ended up sleeping together – I didn't have the willpower to say no. I remember feeling

* You met the amygdala in Chapter 3. Its main job is to identify whether the information it receives is relevant to our survival. The amygdala doesn't stop and think, it just gets you ready to fight back or flee.

very guilty and ashamed of myself afterwards. Even now, years later, I still have that reaction.'

The danger period

As I discussed earlier, giving in to temptation involves different stages, but the most dangerous is the moment when someone finally gives in and crosses the line.

Of course, this is likely to come after a number of smaller decisions on the slippery slope – the phone calls, the burgeoning friendship, time spent together, secrets shared.* But if you've made your way to the bottom of the slope, and the temptation is so close you can smell it, how do you resist? It got me thinking about something I had read many years earlier – the idea that there is a 'danger' period where short-term desire suddenly spikes and overtakes long-term goals or values.

In Mifune's example, she had a long-term aim *not* to get involved with a man who was unavailable. She also had a short-term desire to sleep with a friend of hers – who happened to be engaged. On Monday, Tuesday, Wednesday and Thursday, Mifune's long-term desire was much stronger and she did not want to sleep with him. This was still the case until about 6.00 p.m. on Friday when she started to feel very lonely and was faced with staying alone in a grim and grimy hotel room. She arrived at her friend's house at 7.30 p.m. and then her short-term desire overtook the long-term aim. By 9.00 p.m., they were kissing.† After the

---

* Depending on your definition, this may in itself constitute infidelity – but I'll explore that later in the chapter.

† For the avoidance of doubt, I've completely fabricated all these timings. I would be amazed if, years later, Mifune was able to provide me with such precise times when recounting the story ... But I need specific times to illustrate my point.

event, the long-term desire overtook the short-term and Mifune felt ashamed of what she had done.

A graph might look a little like this:

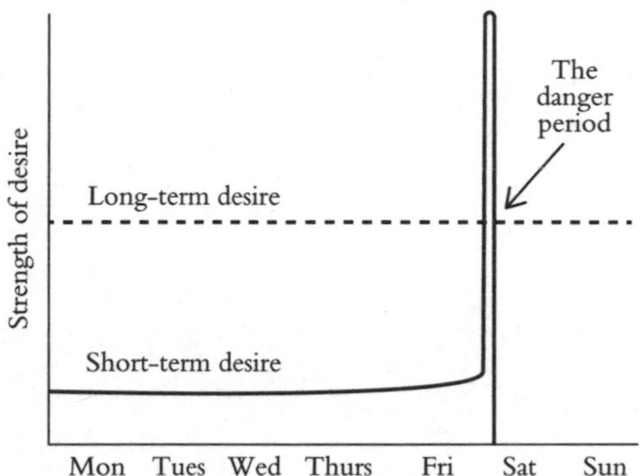

Mifune's long-term desire was the clear winner at all times other than a critical period that started at 6.00 p.m. on Friday. In this high-risk period, the short-term reward spiked above the long-term, undoing many years of restraint and resulting in several years of shame. During this high-risk period, behavioural economist George Ainslie argues that strategy is critical.

Ainslie suggests a number of strategies to help resist the temptation of short-term reward, although I have to say I'm not convinced that all of them will work in the case of infidelity. It's not just marshmallows that Mifune is trying to strategise against; she faces something much more irresistible, powerful and dangerous. She has to strategise to protect herself against succumbing to lust.*

* Infidelity isn't always just about lust – in some cases, it may well be about love. But when people have only just met, given what we explored in Chapter 2, lust is the likely culprit.

Anti-lust strategies

Ainslie's first strategy is something he calls 'external constraints'. In Mifune's example, given that it will be harder to resist temptation once in her friend's house, she could use 'external constraints' to minimise the likelihood of giving into temptation. She could avoid going there in the first place, bring another friend along with her or meet him in a public place.* This is also a good way of limiting the hot triggers of the mental engine, namely, taste, touch, smell, sound and sight. Children who didn't look at the marshmallows or pushed them further away were much better at resisting temptation.

Alternatively, she could get other people to discourage her. Before going to his house, she could call a friend who might remind her of the long-term costs of giving in (or physically prevent Mifune from going round). Or she could completely avoid him altogether.

If Mifune makes it to his house, Ainslie would suggest diverting her attention from temptation. So she could consciously focus her attention on the film they are watching rather than focusing on how attractive her friend is. (Let's hope it's a fascinating film …)

Finally, she could try to make her decision as to whether to sleep with him or not based on a 'universal rule'. A number of philosophers recommend this – Immanuel Kant, for example, believed we should make decisions by 'universals' not 'particulars'. Mifune could make her decision based on a rule ('I don't sleep with men who are in

---

* Given that we know alcohol can make people seem more attractive, and taking drugs or alcohol is often quoted as a reason for 'crossing the line', it might be a good idea to go to a cafe, for example, rather than a bar. And preferably a cafe that's overcrowded and playing terrible music. And she'd be wise not to meet him when ovulating.

relationships') rather than on the circumstances ('I am feeling lonely and this man is very attractive').

You're more likely to stick to a universal rule if it's unambiguous. In Mifune's example, 'not doing *anything* sexual with someone in a relationship' might make an easier rule to stick to compared to one that allows 'everything apart from sex'. The line between sex and nothing is obviously much clearer than the line between sex and 'everything apart from sex'. It's much easier to renegotiate the latter. Mifune might think, 'Well, I've come this far, there is not much difference between kissing him and having sex.' But it's much harder to run that line of argument when your rule is black and white. And, of course, kissing is likely to trigger the brain's engine, making it even harder for the mental brakes to kick in.

It's not just a one-off

Another strategy is 'bundling', by which Ainslie means that we bundle together the long-term rewards (and losses) and then compare them with all the short-term rewards (and losses). The idea is that the collective weight of the long-term rewards is more likely to outweigh the short-term in the danger period.

Amazingly, both human and animal studies suggest that bundling works. In one study, rats were provided with two levers: the left provided sugar water immediately and the right provided the rats with more sugar water – if they could manage a three-second delay. When one dose of sugar water was dispensed, rats chose the immediate (and smaller) reward. But when three doses were dispensed (with some time between each dose), rats opted for the delayed, long-term reward. By bundling sugar water into three doses, researchers managed to change the rats' preferences from short to long term.

Ainslie argues that, as humans, we can bundle choices together simply by using our minds. This essentially means telling yourself that this is not an isolated decision, that how you act now will impact how you act in the future. And there's evidence this works. In one study, students managed to exert more self-control when researchers suggested to them that their current choice might be an indicator of what they would choose in the future. Their decision was no longer a one-off, it was bundled with future behaviour. In other words, they upped the stakes.

In Mifune's example, she could tell herself that giving into temptation at his house would not be a one-off – it might make her more likely to do something similar in the future, transforming the choice into something much larger and, as a result, making the short term less compelling.

The cable car arrived at the destination and Mifune agreed to keep talking to me as we walked into the village. I listened to her through the crunch, crunch, crunch of my boots on the mountain snow. 'I didn't have the willpower that I hoped I would have in that situation. I think there are lots of factors that determine how high your willpower will be. I think the biggest danger can often lie in the things you believe you will not be susceptible to, where you think, "I'm not the sort of person who would do that."'

It was a good point and I agreed with her. I was reminded of a study of 2,025 adults in the US where 88 per cent believed that it was totally unacceptable for a married person to have an affair. I started wondering how many of the 88 per cent of people who strongly disapproved of infidelity thought they would never cheat themselves – but eventually did. Perhaps, then, one of the most important things anyone can do is remember how easy it is to be weak-willed and to recognise that – even if you, and the 'they', disapprove of it* – you still might be susceptible to temptation.

It's easy to underestimate the power and prevalence of infidelity, temptation and lust, but we are all vulnerable (to different degrees). Philosopher Arthur Schopenhauer believed that lust was the ultimate goal of almost all human endeavour, that it impacted the most important affairs, interrupted the most serious business, and confused even the greatest minds (including scholars and statesmen). Countless stories in the press involving high-profile people (as well as the presence of Fortune 500 and governmental email addresses in the Ashley Madison data leak†) suggest he was right.

As we reached Mifune's hotel, she dumped her bag on the snow, folded her arms and said, 'It's important to realise that there are some things that trigger weak willpower in you. I was really lonely. I was feeling really lost and far away from support and companionship – and he represented a lifeline. In my head, I translated love and a loving relationship as a substitute for some sort of solidity. Because I was vulnerable …' She looked down at the fluffy, glistening snow and dug her boots into it. 'Because I was vulnerable, the short term was apparently more important to me than feeling good about myself in the long term when I had realised what I had done.'

How could you, Mum?

'I've never been cheated on. And I've never cheated. But my dad had an affair.' Chris and I sat huddled in the corner

---

* Or perhaps even *because* it's disapproved of – because that makes it more dangerous, which can kick the fight or flight response into action and make someone seem more attractive.

† Who uses their work email address on a cheating website? Seriously?

of a mahogany panelled bar while he took a break from work. He was 19.

'I've seen it happen to somebody that I love and I know damn well what it's like to be cheated on. I know that I would never do that to anybody. It was just seeing my mum in that state … I know that if I cheated on someone else, they'd be in exactly the same position. I've never done it. Never have, never will.

'He told Mum there was a rumour going around about this girl from work. She obviously thought nothing of it – we thought nothing of it. Then we found out six months later. They'd been having marriage problems … they were arguing a lot. Then he told her straight to her face, "I've found somebody else."

'My parents were together for 20 years. They obviously split up – he stayed with this other woman. For Mum, it was 20 years of her life just gone like that. She had all sorts of emotions. She was angry, sad. At first, she found it hard to take for obvious reasons. That was only about two years ago.

'This woman isn't very nice. I'm going to say that because I live with my mum, but she's honestly not a very nice person.

'We don't see him any more. He moved up north 18 months ago. He just upped sticks and left. I don't think it's right that people can get away with it. I don't mean that there should be some sort of punishment, but the way they can just get up, literally, get up and leave their kids and leave everything behind, I just don't think it's right. I've got two brothers, one is six-years-old and he probably won't know who his dad is because of how long ago he left.

'I've seen him once this year. My 15-year-old brother hasn't seen him or spoken to him once since he left. He used to try quite a bit to see us, but it got to a point where we kept saying no. It's sad, but we've got on with our lives and we've thought, "We don't need him any more, we're fine without him." So, we've let it slide.

'I think it would be the same for me whether he'd had a full-blown affair or a one-off thing. But maybe him moving away wouldn't have happened because he wouldn't have been with this woman. I suppose with a one-off, there's less deception. But with a long affair, you've lied for that whole period.

'I'm a bit more wary of people now. Despite the problems my mum and dad had, I never thought this sort of thing would happen. I suppose it's going to be harder for me to trust other people. I'm going to have to be more thoughtful about certain things. I might change as I get older. But for the time being, I won't just let anybody in – I'd have to get to know them first.'

It's really hard to pick apart how infidelity impacts children because it's usually only part of a much bigger picture. In Chris's case, he was faced with a whole host of problems: his parents were arguing a lot and one or both of them may have felt anxious before his dad admitted the affair; he found out that his dad had been lying; and the affair would not only have impacted him, but also his brothers and his mum. Then there was the divorce and its emotional and financial impact (which may have also created more arguments); the emergence of a stepmother (who he knew had been having an affair with his dad); and, finally, the departure of his dad.

I'm going to briefly look at some of these, but will come back to them in more detail in Chapter 13.

Conflict

Conflict can have a big impact on kids – not only does the arguing itself have an impact, but conflict can also influence the way people parent. In one study, mothers in high-conflict marriages were less affectionate, empathetic and warm towards their children, and more rejecting and harsher in their discipline. Fathers tended to be more insensitive and intrusive. Relationship conflict has also been linked with depression, which can limit a parent's ability to be responsive to their kids.

Infidelity

Arguing aside, there's the infidelity itself. As well as impacting the children directly, it's likely to affect the parents themselves (particularly the non-cheating parent), which will in turn impact the kids. Non-cheating partners can experience symptoms similar to those seen in PTSD, and are more likely to be depressed than someone whose partner hasn't cheated. Although it's hard to establish whether infidelity causes depression or if depression somehow makes infidelity more likely, research has found that women who had recently been cheated on were more likely to be depressed than those who hadn't – even when taking into account a lifetime history of depression, their family's history of depression and other negative life events.

Neither parental PTSD nor depression is good for kids. A 20 year study following children whose parents were moderately or severely depressed found that they were three times more likely to suffer from major depression, anxiety disorders or substance abuse than those whose parents weren't depressed.

On top of that, children often experience guilt, anxiety, fear, worry, depression, aggression and shock when faced with parental infidelity. But the impact can vary depending on a child's gender, age and culture – younger children may worry that their parents no longer love them, whereas teenagers are more likely to make moral judgements about the affair. And if a child is forced to provide emotional support to the non-cheating parent, they may feel even more anger or resentment to the person who cheated.

Children can also go on to experience problems when they eventually have adult romantic relationships of their own – particularly when trusting others. In a study of adult children whose parents had been unfaithful, 80 per cent believed their attitudes towards love and relationships were influenced by their parents' infidelity. Seventy per cent said that it had inhibited or reduced their ability to trust their romantic partners.* This was something Chris recognised in himself, saying: 'I never thought this sort of thing would happen … it's going to be harder for me to trust other people.'

Divorce

Infidelity can lead to divorce, which can be bad news for kids.† Children whose parents divorce are more likely to have long-term emotional, social, academic or behavioural problems than those whose parents stay together. Some problems may not appear until later in life, particularly when children become adults or have long-term romantic relationships of their own. In a long-term study that followed more than 17,000 people aged seven to 23, researchers found that the likelihood of having an emotional disorder was 11 per cent for young adults who had experienced parental divorce, compared with 8 per cent for those who hadn't.

This doesn't mean divorce should always be avoided at all costs – a well-managed divorce where a child feels loved and supported by at least one parent (preferably two) is usually much better for everyone involved than struggling on in a high-conflict or abusive marriage.

And in any case, infidelity doesn't always end relationships. Some go on to be very successful after an affair and

---

* Admittedly, we're relying on what people say here – and, as we've already explored, what people say isn't always entirely accurate.

† I'll explore when and why it can be bad for children – together with how to minimise its negative impact – in Chapter 13.

avoid at least some of the negative impact on kids that researchers fear. I spoke to someone in New York who was adamant that his mother's infidelity had not impacted him at all. His name was Orpheus. He was tall, with prominent cheekbones and a tailored petrol blue suit.

'My father was chopping tomatoes in the kitchen and my brother was helping Mum by working out a setting on her phone when he said, "What's this?" and he proceeded to read out a message from a man called Sebastian. Everyone in the kitchen fell silent. Then my father said, "Don't worry, that is just Mummy's boyfriend." It was his way of letting us know that he knew.

'I came home the next day and Mum had a cut on her eye – Dad wasn't a violent man, but they had the mother of all fights and he had hit Mum. My brothers and I were furious, but she said, "You don't know the whole story – I was having an affair" and she started crying at the kitchen sink.'

Listening to Orpheus, I wasn't surprised that his father hit his mother as I'd come across plenty of evidence to suggest that men often get violent when faced with infidelity. Internationally, women are far more likely to be punished for indiscretions than men.* And when psychologists Daly and Wilson researched spousal murder in a variety of societies, they found that the story was basically the same: most happened as a result of the husband's jealous, possessive and violent response to his wife's (real or imagined) infidelity.

Orpheus leant back into the leather chair and took a sip of his cocktail. 'A few months later, I met Sebastian. He came round and we sat by the pool together and drank sparkling water. I wasn't sure whether the affair had actually

---

* From beating among the Huichol of western Mexico to thrashing or even death among Trobriand Islanders in Papua New Guinea.

ended or whether they were just pretending that it had. I didn't know and I didn't care. I could see why Mum fancied him – like her, he was American and together they talked about books and theatre. My father was Austrian and had no interest in either.

'The affair ran its course. When it ended, Mum would smirk if I mentioned it. She knew I wouldn't particularly judge her. To Mum and Dad's credit, they now have a really good marriage. They are in their 70s, very good friends and really happy. I can see they will be together forever.

'My father has very high self-esteem, he's not jealous and he's very practical. He didn't want a divorce because he and Mum have children together, she's really good company and was never going to leave him for Sebastian. Dad may also have been partially to blame, they may not have been having sex very much.'

Orpheus took another sip of his drink and leant forward. 'A few years later, when I was 14, Mum and I were visiting family in Paris when an old family friend came over for dinner. There was talk of him staying on the sofa, so I naively said, "Oh, I'll help you make a bed," and then we all went into different rooms to go to sleep.

'The week after, I was on the phone to my dad catching up and I said, "Oh yes, and we saw Jeremy, he stayed over." Mum kept gesturing for me to be quiet, so when I put the phone down, I said, "What was all that about? Why can't I tell Dad that Jeremy stayed over?" "Because I slept with him," she said. I remember not batting an eyelid.

'It's such a cliché when you see children on TV dramas saying, "How could you, Mum? How could you do that to Dad? I am never going to talk to you ever again." It wasn't a life-shattering event. It was just another of Mum's admirers. She left a trail of broken hearts behind her and then slightly played up to it and enjoyed it – not in a malicious way.

'The flip side is that she'll acknowledge that she herself is responsible for not finding true love. Although what she has now looks like true love to a lot of people – perhaps because it transcends sex. Now she talks a lot about companionship and says, "Oh darling, it's so lovely when you get to 60 because you don't want to have sex any more." But you need sex when you're young.'

Don't talk about it, but don't lie …

I was struck by the very different responses Chris and Orpheus had to their parents' infidelity. Perhaps it was unsurprising given the very different circumstances and consequences – one resulted in the break-up of a marriage, the other didn't. But a number of other factors, including their experiences until that point, who they both were as people as well as their attachment styles, will also have played a role in how they dealt with it.

Their different responses were particularly clear in their parting thoughts. 'Don't lie,' said Chris. 'Lying doesn't get you anywhere. They will always find out one way or another. Lying's an easy way out but try and avoid it if possible. Don't lie to them. Or try not to.'

Orpheus, on the other hand, took a different view. 'Ultimately, I don't think sexual attraction is the most important element of a romantic relationship. If one of you does cheat, as long as it's not happening all the time or it's not a huge love affair that is threatening the marriage, I don't think you should talk about it. If it happens, it happens. Occasionally in life, everyone slips up* and it's naïve to think that they won't.'

---

* The stats we explored in Chapter 1 suggest that, in the context of cheating, he might be right – many (though admittedly not *all*) cheat.

I admit that I screwed up

I met Andrea and Dan on a boat trip down Manawatu River in New Zealand. They asked me to take a photograph of them and we got chatting. They were 22-years-old and in a long-term relationship.

Andrea: 'I've been cheated on. Once. I wouldn't cheat. If I'm going to commit to somebody, I'm going to commit to them. If I don't want to be with them, I'll break up with them. If I look at someone else and think, "You know what? I really want to sleep with this person," then I clearly don't want to be with my boyfriend.'

Dan: 'Imagine that you are dating someone for a year, then you meet someone who you totally click with. It happens.'

Andrea: 'But you shouldn't go looking for it.'

Dan: 'Absolutely not, but if you're in an office, a bar and you completely hit it off with this person like you've never hit it off with anyone in your life, I think that's something you have to think about.'

Andrea: 'But at the same time, don't you have to be honest with the person who you're in a relationship with?'

Dan: 'Yes, but I can still connect with someone without the thought of us sleeping together that night.'

Andrea: 'But you're connecting with someone else instead of the person that you are in a relationship with, the person that you said you were going to connect with.'

Dan: 'We have different meanings of connection. You think of connection as flirtation, two people who are going to get it on.'

Andrea: 'No, no, not at all – I think of it as an intellectual connection, being with someone and wanting to talk to them, not necessarily wanting to sleep with them.'

Dan: 'So what would be wrong about it?'

Andrea: 'As long as you're honest with it, it doesn't matter.'

Dan: 'But where is the dishonesty in it? You're just having a conversation with somebody.'

Andrea: 'But if you know as a person that you have a connection or any sort of feelings with this person, you should say something about it to your partner.'

Dan: 'What about your mum and her best friend Jim. They are great, great, great friends. They have known each other for decades. They probably talk to each other in a different way to which he talks to any other person – any other woman – on this planet. But they're not going to sleep with each other. It's the same as meeting someone in a bar – two people can have a conversation and connect – you just assume that if you talk to somebody, you're going to sleep with them.'

Andrea: 'Perhaps that is because I've been cheated on – and it all began with a conversation.'

Dan: 'Not all men talk to women just to sleep with them, and vice versa. In a business environment, women and men work together every day. Men and women can be friends without having a sexual connection. And there are a lot of good-looking people on this planet. You're allowed to say if someone is a good-looking individual without wanting to sleep with them.'

Andrea: 'There's a fine line though.'

Dan: 'Yes, but you are allowed to say, "Jude Law is a good-looking guy" – there is nothing wrong with that. When you're in a relationship, I think you have to give the benefit of the doubt to your partner. As a general rule, people value their time, so you have to ask, "Why would they be investing their time in our relationship if they are after somebody someone else?"'

Andrea: 'But we've been in a relationship for years, I appreciate that you have invested time, but I'm still messed up because you cheated on me.'

Dan: 'I admit that I screwed up, but I don't think it's fair that, in random arguments, you bring up some of the darkest moments in our relationship. It's petty and disrespectful to still be bringing it up two years later. That girl is so far behind us, we've gone so many places, we've done so many things, why do you feel it is necessary to bring that up? Has nothing in the last two years meant anything? Do you not sincerely think that two years later, it means nothing?'

Andrea: 'In certain situations, it feels relevant to bring it up.'

Dan: 'There is nothing that will make you forget it. You will hold it against me forever.'

Andrea: 'I probably will because I'm too insecure. You're too charming, women love you and it creates problems. You do it naturally.'

Dan: 'As a result, I feel I'm not allowed to talk to women, ever. And in my office, I can't avoid working with women. Not just in the office – as there are more women on this planet than men, I'm going to end up talking to them at some point.'

Andrea: 'As long as you realise that a lot of women that you meet would like to be with you.'

Dan: 'I really think it's you who has to realise that I have no intention to be with other women.'

Andrea: 'It's not you that is the problem – it's the other women.'

Dan: 'Yet I get the backlash. If you knew that I loved you, you would never feel jealous and you would never feel the need to act out when that happens. Why would I spend the last two years with you if I didn't love you or care about you?'

Andrea and Dan's interview was probably one of my most awkward – it was more of an overheard argument than an interview.* It reminded me of the question I mentioned in Chapter 1 about what exactly infidelity is. Is it sex? Kissing? An emotional connection? Or something in between?

Most studies I've looked at assume that infidelity involves some form of sexual contact, but some psychologists argue that it doesn't. Intimate romantic relationships with other people can be very harmful, even if they don't involve anything physical. And cybersex, where people only ever meet via a screen, can feel just as much of a betrayal as sexual contact and can be just as damaging to a relationship. In one study, more than 50 per cent of people who engaged in cybersex lost interest in sex with their spouse.

Psychologist Shirley Glass, who spent her life researching indiscretion, saw infidelity as a secret sexual, romantic or emotional involvement that violated the commitment to an exclusive relationship. For the sake of simplicity, I define it as some form of sexual contact (usually sex), mainly because that's how it is described by most of the research. But there are no clear lines on this, and given the choice, some might prefer their partner to have a one-night stand than a purely emotional affair lasting many years. It also depends on what (if anything) has been agreed, promised or assumed by the partners – even if it isn't technically cheating, it can still be a betrayal.†

You'll hold it against me forever

Aside from questions about precisely what infidelity is, what really struck me was the huge impact Dan's infidelity had had on Andrea – so much so that years later, she and

---

* I should add that they both gave me permission to use the recording for this book.

† Betrayal doesn't require any sexual contact. (Betrayal – in this context – means being disloyal or disappointing someone's hopes or expectations.)

the relationship were still very wounded by it. I didn't ask what happened at the time, so I don't know whether she fully processed how she felt, whether they talked about it or whether he showed remorse. But two years later, she continued to punish him and didn't seem able or willing to forgive. And they both seemed to recognise this when Dan said, 'You will hold it against me forever' and Andrea said, 'I probably will'.

There's no avoiding it. Infidelity can be very damaging. It breaks trust and disrupts assumptions. Most people (or at least those in straight relationships) assume their partner won't have sex, or an emotional or romantic relationship, with anyone else.* Infidelity shatters that.

As I mentioned, people who've been cheated on often experience symptoms of depression, anxiety and rage. Even when taking into account both their own and their family's lifetime history of depression and other negative life events, women who had recently been cheated on were more likely to be depressed than those who hadn't.

Cheating is notoriously difficult to resolve in therapy. In one study, couple therapists ranked it as the third most difficult problem to treat – after a 'lack of loving feelings' and alcoholism. And when researchers followed couples for five years after therapy, they found that relationships where one or both partners had cheated were more than three times likely to end than those where both were faithful.

But it's not all doom and gloom. Not all couples divorce. Nor are the long-term effects always negative. For a

* Most – but not all … And research suggests that it's less of an assumption for some members of the LGBTQ+ community, but I'll look into this in the next chapter. In any case, in a US study, 92 per cent of lesbians and 82 per cent of gay men said they wanted to be in a monogamous relationship after the age of 30, so whether or not it's assumed, it seems to be something that many hope to achieve in the long term.

minority of couples, an affair can galvanise them into tackling problems that were gnawing away, and they can come out the other side with a better, happier and more intimate relationship. But for that to happen, the person who's been cheated on has to find a way to forgive the cheater (just as the cheater has to find a way to forgive themselves). And that's not easy.

Forgiveness

Philosopher Margaret Holmgren argues that in order to genuinely forgive someone, you usually need to go through a process of responding to what they've done wrong. This involves:

1 Recovering your self-esteem (which is important because, as philosopher Jeffrie Murphy argues, betrayals can communicate things, they can be a way of saying, 'I count but you don't').
2 Recognising that your partner's action was wrong – and why it was wrong.
3 Acknowledging your feelings and experiencing them in full.
4 Sharing your beliefs and feelings with your partner.
5 Assessing your partner for who they are (not who you may want or demonise them to be), examining their attitudes and behaviour patterns, and taking steps to avoid them causing you more harm. This might mean ending the relationship or changing its terms.
6 Deciding whether you want any form of reparation.

Once you do this, Holmgren argues that forgiveness is always a good thing, regardless of what you've suffered and whether your partner repents. But forgiving someone doesn't mean staying in a relationship with them. While she would advise victims of domestic abuse to forgive their abusers, she wouldn't necessarily recommend they stay in the relationship (nor would I). Instead, they might try to understand why their partner is abusive and find a way to overcome negative feelings towards them, but at the same time, decide that they can't live like this and walk away.

The ultimate aim of forgiveness is to try to process your emotions and reach a sense of acceptance. Philosopher Bishop Butler, one of the first thinkers on forgiveness, describes it as an attitude of real goodwill towards the offender. But that's no easy task when someone has breached your trust and caused you (and perhaps also your children) enormous pain.

Holmgren suggests that the way to do it is to see your partner as a human, who is flawed and vulnerable to error as all humans are, and to think about the bigger picture – particularly because the people who go way off track are often those who've had more difficult pasts.*

In order to let go of negative feelings, she argues that you have to recognise, process and communicate them to your partner. But processing emotional pain is often much harder than trying to avoid it by, say, burying it. Maybe that's why Ghandi thought weak people could never forgive and saw forgiveness as an attribute of the strong. (I'd argue that this strength is particularly important when children are involved.)

Cold water splashed up onto the deck of the boat, creating a small rainbow in its wake. Dan repeated his last question, raising his voice slightly. 'Why would I spend the last two years with you if I didn't love you or care about you?'

Andrea watched the waves undulating below. 'You are inconsiderate.'

---

* And might also be having a difficult present.

'You just called me names and did not consider my question. Why would I have been with you this long if I wasn't invested in this?'

'I didn't say you weren't invested. I said that I thought you were inconsiderate.'

'You saying I'm inconsiderate pales into comparison when you bring up dirt from two years ago.' He tried to make eye contact, but failed. 'That's not kind, that's not what people who love each other do.'

CHAPTER TEN

Sex, Swans and Swingers

So far, in talking about infidelity, I've made a big assumption: that romantic relationships are or should be monogamous.* It's not wildly unreasonable to make that assumption because in Western societies, most people are in exclusive relationships; or at least they think they are, even when one or both people are actually unfaithful.† Moreover, this two-partner approach is often seen as the best way to have a relationship in the West – it's presumed to be the normal and natural way for humans to pair off. ‡ But maybe it's not.

What animals get up to

Very few animals are sexually monogamous and DNA testing has shown that many of those we thought were, aren't. Swans, for example, can mate for years, even for life. But in any given clutch of eggs, 40 per cent contain at least one fathered by a different male.

---

* I'm talking about relationships generally – marriage is a stickier topic because legal and religious factors come into play.

† Biologists would describe this monogamous-but-cheating scenario as 'socially monogamous' but not 'sexually monogamous'. Social monogamy means having an intense emotional and physical bond with one person, often making a home and family with them, whereas sexual monogamy means sexual exclusivity. When people (other than biologists) talk about 'monogamy', they're usually referring to both social and sexual forms.

‡ Psychologists call this way of thinking 'mononormativity' (as in, the normalisation of monogamy) – or, in the case of straight monogamy, 'heteromononormativity'.

Out of 4,000 mammal species, only a few dozen are monogamous, and of the world's 300–400 primates, only nine are, which works out as 1–3 per cent.* So, not many, then. And our closest living relative, the bonobo,† isn't into monogamy – instead bonobos have sex to resolve disputes, practice, say thank you for food, comfort each other and reinforce social bonds. (Which makes me wonder if there's anything they don't use sex for?)

Of those that do manage monogamy, it's a lot easier than it is for humans. Californian mice, for example, reject the chance to mate with anyone other than their partner, instead opting for lifelong monogamy. But then they don't live for more than two years. I'm pretty sure infidelity rates would drop dramatically if humans didn't live beyond being a toddler.

But what about humans?

Although typical in Western societies, monogamy isn't actually the norm in humans either. For decades, anthropologists have studied a variety of non-monogamous societies around the world, societies that see things very differently to most Westerners. Anthropologist Geoffrey Gorer described how the Lepchas of the Himalaya see sexual experience as almost divorced from emotion. They think it's an enjoyable and amusing experience, as necessary as food and drink, and it doesn't matter who gives it to you, as long as you get it. When he stayed in the village of Lingthem, Gorer was surprised to see that it was home to

---

* In case you're wondering which of the mammals have succeeded in monogamy, you're looking at a few species of bats, some foxes and monkeys, a giant otter, a beaver, a few seals, a sprinkling of mice, rats and rodents, and a couple of small antelopes.

† Full disclosure: some argue that bonobos *and* chimps are the closest living relatives to humans.

two 60-year-old women whose invitations for sex were never refused – and many of the men who slept with them had young and pretty wives.

In China, the Mosuo have an arrangement they call *sese*, which means 'walking'. When a girl turns 13 or 14, she's given her own bedroom that opens on to the courtyard of the house and directly on to the street, and she has complete choice over who she allows in. The only rule is that they must leave by sunrise. She can have a different lover every night, or multiple lovers in the same night if she chooses. There's no expectation of commitment, and any child she has is raised with the help of her family and the community.

Over in Peru and Brazil, the Kulina have an intriguing pre-hunting ritual. A group of village women go from home to home, singing to the men of each household. As they reach each house, at least one steps forward to knock on the entrance with a stick. Whoever does so sleeps with the men of the house that night – but only if they are successful in the hunt. (I should also add that women aren't allowed to choose their own husband.) But that doesn't lead to as much competition as you might think. Once the men have finished hunting, they meet up outside the village and redistribute what they've caught … This means that every man goes back with meat, and so gets to have sex (with someone who isn't his wife) on their return.

Though it might seem surprising, a number of societies around the world believe that a child can have more than one biological father.* Many of these believe that extra semen helps foetuses grow, and in some cases, like the Aché

* The technical term is 'partible paternity'. Cultures that believe in it include the Aché of Paraguay; the Barí of Venezuela; the Kulina of Brazil and Peru; the Matis in Brazil; the Pioroa of Venezuela; the Secoya of Ecuador, Colombia and Peru; the Warao of Venezuela, Trinidad and Tobago, Guyana and Suriname; and the Yanomami of Venezuela and Brazil.

of Paraguay, the second father is expected to give the child food and get more involved in its life if the first father dies.

This concept of multiple fathers is so accepted that the Aché actually have different words for different types: the father who put it in (*miare*); the fathers who mixed it (*peroare*); those who spilt it out (*momboare*); and those who provided the child's essence (*bykuare*).* In cultures like these, non-monogamy isn't just tolerated, it's encouraged, because people believe that it will result in stronger children. And there's evidence to suggest that children with more than one father are likely to live longer: anthropologist Stephen Beckerman found that only 64 per cent of Barí children with one father survived to 15-years-old, compared with 80 per cent who had two.

Let's not forget polygamy (the practice of having more than one spouse – although it's almost always polygyny, which is having more than one wife). Of the 160 countries examined by the OECD† in 2009, polygamy was legal or accepted in 33 of them, and partially legal or accepted in a further 41 (mostly in Africa and Asia).‡

For some, 'partial' acceptance is only a relatively new approach and polygamy had been accepted for years until recent reforms. In Morocco, for example, polygamy was completely legal until the Family Code 2004 (known as

* In case you're wondering how someone goes about giving a child their 'essence' – it seems to have a lot to do with providing the mother with meat to eat …

† There are 195 countries in the world, but the OECD only looked at 160. This is just one of the many sets of statistics on polygamy. And it's hard to say if it's the *best*… That's because there are so many ways to define and categorise what polygamy actually is, and this makes it difficult to establish whether one survey is closer to lived life than another.

‡ And that's not the full picture – unofficial polygamy also happens beyond these continents, although it's hard to know exactly how many people do it.

the *Moudawana*) made marrying more than one woman more difficult (but not impossible). A man is now required to get permission from his first wife before marrying a second, as well as making sure the second is aware of the first, proving to the court that he can financially support both wives and any children equally, and that he has exceptional circumstances.*

'In Morocco, most men have only one woman. Rarely, very rarely, you might find, maybe, one single man in a neighbourhood who married two women – but only if there is a problem with the first wife, if she can't have children, if she is seriously unwell.' I met Ahmed just outside Marrakesh. He was in his 40s, tall and wore an elegant white cotton djellaba. He spoke slowly and carefully with a deep husky voice.

'He can't just throw the first wife out because most women don't work. He can't just divorce her. But he can marry another woman as well. You can marry up to four women. But in Morocco, you must make it legal with the women by getting their permission. And why would a woman give a man permission to marry another woman just because he feels like it? You can't marry four women just like that.'

Not so normal?

If only one or two countries allowed polygamy, or one or two tribes were non-monogamous, we could perhaps write them off as outliers, returning to Western monogamy assured that it's the natural way for humans to have relationships. Yet 44 per cent of countries examined by the OECD accept polygamy to some extent, and non-monogamy is common among people who live thousands of kilometres

* Chapter II, Articles 40 & 41. Although, as with any country, the efficacy of the legislation depends on how robust the rule of law is.

apart, speak unrelated languages and show no sign of having been in contact with each other for centuries.

Add to that the overwhelming non-monogamy of the animal kingdom (including among our closest relatives), as well as the high rates of infidelity in many 'monogamous' cultures, and monogamy seems to be less 'natural' or 'universal' as many might believe.

Non-monogamy in the West

Non-monogamy describes a number of very different arrangements, including open relationships, swinging, polyamory and polygamy.* Academics have given this the rather catchy label of 'consensual non-monogamy', but that's a whopping nine syllables, so I'm going to steer clear of it. Instead, when I use the term 'non-monogamy', I'm going to assume that everyone consents to what is going on.

The evidence seems to suggest that even though most people in the West practise monogamy, globally, the norm is actually non-monogamy in both animals and humans. But it's also more popular in the West than some might realise. In two studies totalling 8,718 single people in the US, for example, more than 21 per cent had tried it out.†

---

* I'll explain these in a moment, but just so we're clear: polyamory involves multiple emotional relationships that are often sexual (but not always) and polygamy means having more than one spouse. Some academics also include another category of non-monogamy: gay non-monogamy. But most include this within open relationships and I'm going to do the same.

† On the one hand, this percentage might be too high – only single people were questioned, and they might be more permissive or non-traditional than people in long-term relationships. On the other hand, as with infidelity, some may try to keep their non-monogamy quiet for fear of judgement or stigma – and if so, the percentage could be higher. (That said, the questionnaires were completed anonymously online, which reduces – but doesn't eradicate – that risk.)

Men were more likely to have tried it than women, as were gay men, lesbians or bisexuals compared with straight people. Interestingly, age, education level, income, religion, region and race had no impact; so thinking that people who try it are richer, poorer, older or younger than you isn't justified. There is no singular type of person who tries it. Some people know they want more than one intimate partner from the start, whereas others become curious after years of monogamy.

The rules for non-monogamy are open for discussion and people negotiate whatever works for them. That means they can be as individual as the people involved and this makes it difficult to generalise.

There isn't much research in this area and much of what does exist relies on recruiting people from internet forums. But that misses out heaps of people who don't use forums, as well as those who might not want to volunteer to take part in studies (say, where non-monogamy isn't working so well for them). It also omits the people who tried it and didn't like the experience for whatever reason. This means I can't make valid sweeping generalisations. But I can use the evidence to give you an idea, however limited, of how non-monogamy works for some people.

Open relationships

People in open relationships often think of there being different types of monogamy and couples are expected to be emotionally, but not sexually, monogamous.* Rules tend to play a big role and can include: whether anything is off

* Unhelpfully, the phrase 'open relationship' is used in various ways and can also be used to describe all non-monogamous relationships – but I'm going to ignore that definition and stick with the more specific meaning of the phrase.

limits; whether you can sleep with any gender; how you're going to protect against STIs; whether you can meet the same person more than once; whether you can sleep with someone you both know; when and where you can have sex; and whether you should talk about it afterwards or not.

Studies of gay couples have found very different approaches to each of these rules. For example, when it came to talking about what they got up to, some operated a 'don't ask, don't tell' policy, whereas others found that discussing liaisons in minute detail helped them maintain trust while revitalising their sex life.

Orpheus* explained: 'My friends who have open relationships have very strict rules and some only play when they are together. But if you don't do it together, then do you tell each other? I'm not sure I could manage an open relationship, but if I ever tried, my rule would be, "I don't want to know – just don't sleep with any of my friends." Perhaps I would always wonder to what extent are they looking for it and to what extent is it coming to them?'

Swinging

Swinging is similar to open relationships in that people can have sex (or at least mess around) with others, but emotional involvement is off the cards.† But a key difference is that there's much more of a social structure around it. Swingers treat sex a bit like any other social activity that can be experienced as a couple – and the key point is that they experience it *together.* (Single people can also swing, but it's rare. So rare that a single female looking to swing is known as a 'unicorn'.)

---

*You met Orpheus in Chapter 9.

† I have read accounts of swingers who've developed intimate relationships with the couples they swing with, but on the whole, the focus seems to be more on sex than emotion.

Psychologist and swinger Dee McDonald explains that, in her experience, many who attend swingers' parties make sure that they turn up and go home together *as a couple* (even if they have sex without each other when they're there). And swingers don't always have sex. Some do what's known as 'soft swinging', where they agree not to go all the way – while others agree that sex is fine, but kissing or cuddling is off limits.

It's hard to know exactly how many people swing. Research carried out by the North American Swing Club Association (NASCA) found that 15 per cent of couples in the US have tried it, but research that *hasn't* been carried out by NASCA has come up with much lower percentages, ranging from 1–4 per cent. Whatever the numbers, it's something that happens around the world – there are swingers' clubs all over, including in Canada, England, France, Germany and Japan, as well as in almost every state of the US.

A recent study of 1,376 swingers found that most were 30–50 years old, had been in their relationship for over 20 years and went swinging 4–5 times a year. Two-thirds had a degree or some form of postgrad, a third were either business owners or professionals (including lawyers, accountants, doctors and engineers) and most had a household income of $70,000–$200,000 (approximately £50,000–£150,000). Most men were straight, most women were bisexual, and only 1 per cent of men and women were gay. Compared with women, men were more likely to suggest trying it out in the first place, usually so that they could fulfil their fantasies or have some variety in their sex lives.

Polyamory

Polyamory involves multiple emotional relationships that are often sexual (but not always). Theoretically, there's no limit to how many relationships a polyamorist might have, but in a small US study of 343 polyamorists, most (42 per cent) had two. Twenty-five per cent only had one partner, 17 per cent had three and 15 per cent had four.*

I spoke to Nora as she ate her lunch (a cheese and chutney sandwich) and sunshine trickled through the sessile oak trees outside. She was tall, slim and Irish, with a boyish haircut. 'Some people will have a primary partner, the partner that means the most to them, that they spend perhaps the most time with or they feel the most love for – and then they will have other partners who are secondary to that. So, there is a bit of a pecking order.

'I have been … someone's secondary, or someone's person that they only saw now and again, and actually, that suited me really well. So, it didn't bother me at all because it was convenient to my lifestyle. But … when I was in my first poly relationship, the people I was seeing weren't always happy that they were secondary. So, it depends. It is all circumstantial and all situational.

'And there are other relationships where everyone is treated equally. That was what my most recent poly relationship was like – the person I was with had another partner and he treated us both with the same kind of level of equality, the same level of respect, the same level of attention … There is no set way of doing it.

'Lots of people read books on it … and they think "there is a way you have to do this". That is not the case at all. It is a very personal thing to negotiate; it depends on the people involved and there are no set rules for it.'

In the polyamory study I mentioned above, 96 per cent of people, regardless of gender or sexual orientation, had

---

* Some polyamorists argue that even this sort of categorising of polyamorous relationships imposes limits that are symptomatic of a monogamous culture. But I think it's worth a mention because it gives you an idea of the variety of arrangements.

some kind of agreement in place. Most (65 per cent) went for verbal agreements rather than writing anything down, with some banning activities like holding hands, spending the night with someone else, having sex and falling in love. That said, these were still allowed by the vast majority (82 per cent).

'Our rules were complete and open transparency,' explained Oscar, a photographer from Cape Town, South Africa. 'And safe sex. I was sleeping with one girl without a condom, but she was the only one. If someone trusts you, they are giving you their health and their life – so if you tell them you are going to have safe sex, you have got to abide by that. If anyone wanted to sleep with me without a condom, I would make sure that we both went to the sexual health clinic together and got tested first.'

And then there are those who deliberately resist having any rules, opting instead for what they call 'freedom-from-contract'. Except there still seems to be some sort of rule or agreement, it's just very short (and hard to enforce). Even if the agreement is to emphasise communication or self-awareness, or to make sure that the primary couple is the main couple, the lawyer in me still sees that as an agreement (and, of course, one that can be broken).

Why bother?

If you can pull it off, non-monogamy can mean having your cake and eating it (or, as some polyamorists put it, 'having your Jake and Edith'): you can have the excitement of getting together with people you find attractive at the same time as retaining the intimacy and security of your relationship.

Research into gay couples, together with a study of 1,093 people in poly relationships, found that non-monogamy didn't impact commitment. (Although in both cases, this relied on what people said about their relationships and, as we know, we can't always rely on what people say … Plus, in the polyamory study, researchers only heard one side of the story.)

Non-monogamy can also re-energise existing relationships: some say swinging improves their sex life at home, while some polyamorists think that 'new relationship energy' spills over into their existing relationship(s). I wondered whether this was down to lust, which was surely at play in both cases. Perhaps lust is almost contagious and capable of spreading to every relationship in the mix – just as narcotics might make everything seem more exciting.

But, as with drugs, lust isn't always a good thing, and some say new relationship energy has the opposite effect, pulling them away from existing relationships in the excitement of the new. And it's not always easy to be the person who feels left behind, which might be why jealousy and insecurity come up in a lot of the research. As I said earlier on, non-monogamy is very individual, so it's impossible to generalise. Non-monogamy can have ripple effects, but they're not always for the better.

Surely one of the biggest advantages is the honesty that, if the infidelity statistics are right, doesn't always come with monogamy. For some people, it's not the sex itself that's problematic – it's the deception. And there's no need for deception if you're allowed to do it. Nora explained: 'That is the main reason that I wanted to be polyamorous … because you are as open as possible and there is no subterfuge, there is no going behind anyone's back, there is no betrayal. Everything is out in the open, so you can make more informed decisions …

'If I had known that my ex-husband was cheating on me, for example … if I had known when that started, then I could have made the informed choice to walk away or kick him out or whatever. But I didn't have that and I think that is what I like about the poly set up … its honesty … You find out things are going to change and you can deal with them.'

(Dis)honesty and other downsides

Despite the pros, non-monogamy isn't all fun and sex games. When marriage counsellors from across the US were asked why clients of theirs had decided to stop swinging, reasons included: jealousy, guilt, threat to the relationship, boredom, disappointment and getting emotionally involved with the people they were swinging with.

In a study of gay couples, many of those in open relationships admitted to feeling excluded, insecure, jealous or resentful. One couple struggled so much with outside sex that they broke up before the study finished. And a lot of the monogamous couples had tried non-monogamy, but decided it wasn't for them.

A number of studies found that a major fear of non-monogamists is the risk of contracting STIs. Of course, this can be minimised if everyone has safe sex. But most people don't when sleeping with their main partner (if they have one), which means that if one person has unprotected sex, it puts both at risk (and maybe others too in poly relationships). One way around this is to have a 'safe agreement' where people promise to have safe sex if they sleep with anyone else (as Oscar did). But that only works if everyone actually sticks to it. A study of 285 gay men in steady relationships found that 10 per cent had broken their safe agreements – when only 12 per cent had one in the first place.

Monogamy isn't necessarily safer, though. If you stayed together with the first person you slept with, you'd obviously reduce the STI risk, but that's not how monogamy works most of the time. Instead, people usually move from one relationship to another, then another, then another … often without having safe sex, so risking infection each time. Then throw in some unprotected infidelity and the risks get even higher.*

Perhaps the most iniquitous of non-monogamy's downsides is stigma. It's such a problem that, in a survey of more than 4,000 polyamorists, 29 per cent said they experienced discrimination – that's more than twice as many African Americans who feel discriminated against (13 per cent).†

People often view non-monogamy as an excuse to cheat, seeing these relationships as unsustainable or not as good as opting for monogamy.‡ Nora explained: 'When I split up with my recent partner, I did have at least two people say, "Oh well, of course it wasn't going to work, it is because you are polyamorous" – as if "normal" relationships don't end. There is the perception that we are all hypersexual or something, I don't know. Or we are greedy or we can't decide, or we are flaky or in some way addicted to love or sex. I think that is pretty unfair.'

It's not just polyamorists that get a bad rap. When social psychologist Richard Jenks investigated perceptions of swingers, he found that they were, at best, seen as unhappy individuals looking to fix unhappy relationships, or, at worst, deviant, socio-pathological individuals in need of counselling (ironic given quite how popular it is to watch

---

* When I spoke to a doctor who ran an STI clinic about this, she told me that she often saw people coming in because their partner had had unsafe sex with someone else – in some cases contracting HIV and then transmitting it to them. She also said that she'd heard multiple stories from gay men attending her clinic of straight, married men approaching them for sex in public places – but not having with them, or thinking to use, any form of STI protection.

† These statistics come from the General Social Survey conducted from 1972–2012, which had a total of 57,061 participants aged 18–99.

‡ There's not enough evidence to establish whether monogamous relationships do actually last longer than non-monogamous ones – but plenty of non-monogamous relationships are long term. Two studies of 343 polyamorists and 1,376 swingers found that a third and more than half respectively had been in a relationship for over a decade.

other people having sex on a screen*). And swingers are clearly aware of this: in a number of studies, those questioned said they worried about being found out.

The sad thing is that honesty is often cited as a key reason for being non-monogamous: people would rather be honest with their partner(s) about their desires than quash them or go behind their back. Yet while non-monogamy might mean that you can be honest within your relationship(s), it can often result in dishonesty to the outside world – because to be open about it would result in steely stigma.

'I know people who are very, very quiet about being poly because their job is at risk, their reputation is at risk and so they don't tell people … they keep it hidden,' explained Nora. And this deception can put a strain on relationships, in part because deception costs energy and is stressful, but also because it can feel like you're devaluing the relationship by not being open about it. A study of 305 lesbians found that whether or not they were honest to others about their sexual orientation had a big impact on their relationship satisfaction.

Stigma also has another unwelcome consequence. In an attempt to protect them from it, or from having to lie or be secretive themselves, some non-monogamous parents lie to (or at least withhold the truth from) their children. In a small study of 71 people in open relationships, a quarter said they would never tell their children about their non-monogamy. And of the kids of poly families who do know what's going on, they keep quiet, they sidestep, they're secretive. Elisabeth Sheff spent 15 years interviewing 206 people in poly families, including 37 kids. She found that when teachers and coaches mistook a child's poly family as two divorced families that got on well, the child usually didn't correct them. And in order to sidestep having

* In case you've forgotten the statistics I mentioned in Chapter 7 – one porn site reported having 81 million *daily* visits in 2017.

to explain who various members of their poly family were, teens and tweens[*] met friends in public until they felt they could be honest about their poly family.

Whenever there are rules, they can be broken

'I was betrayed by my husband,' said Nora, adjusting the sleeves of her shirt and clasping her hands. 'I don't think I could go through infidelity again. It would destroy me. It had such an awful, awful impact on my life. Maybe … I am a bit worried about the potential for cheating, so I look for a relationship where it is not going to happen.'

As we came to the end of the interview, I thanked Nora for her time and honesty, particularly given that many non-monogamists prefer not to discuss their relationship arrangements for fear of judgement. And as I walked away from our conversation, I wondered whether any relationship arrangement, whether monogamous or non-monogamous, could protect against betrayal.

It might be less likely if non-monogamists really do value honesty and self-awareness more than most. But even honest and self-aware people make mistakes, particularly when under the potent influence of lust.[†] And when

---

[*] In case you're not familiar with the term 'tweens', it generally refers to a child aged around 9–12.

[†] And there's plenty of evidence that people in non-monogamous relationships break the rules. In two US studies, more than half of the 39 gay couples interviewed had broken agreements they'd reached about sex with others, and 18 per cent of polyamorists admitted to cheating. The percentage is probably higher because, as with infidelity stats, people don't want to admit because of stigma or embarrassment, but also because they want to distinguish non-monogamy from situations where people are monogamous but cheat. Plus, people were asked if they had 'cheated', but polyamorists don't tend to use that word – they use 'rule-breaking' instead – and so may not have admitted it.

there are more people involved, betrayal becomes more complicated. So, at best, it reduces risk, and, at worst, it makes betrayal even messier.

Wherever there are rules, they can be broken. And, as we explored in Chapter 1, it's the very act of breaking rules and agreements that creates excitement and increases attraction. Remember the wobbly bridge effect? Doing what you're allowed to do because you've negotiated it won't have the same thrill as doing something you're not allowed to do. So, if it's immunity from betrayal you're after, the only sure-fire way to get it is to live in isolation, free from human weakness.

As I delved into research from societies around the world, I found myself coming back to what felt like an inescapable contradiction in romantic relationships: as humans, we often want both safety and excitement, independence and unity – and we want them all at the same time. Monogamy doesn't solve this problem. Neither does non-monogamy. And I suspect that's because the problem isn't solvable.*

* Well, it might be … Psychoanalyst Erich Fromm argues that a *truly* intimate relationship, where you both share the entirety of your being, won't get boring because humans are so incredibly complex – and we can never fully understand each other.

CHAPTER ELEVEN

The Good Fight

'Sometimes you need to take that argument to its highest level and then literally burn that out. From my experience, a lot of our arguments are about behaviour and trying to change behaviour.' Llanzo was Scottish-Jamaican, married and in his 50s. Our interview was punctuated by his rambunctious laugh, which echoed round the empty exercise room we had snuck into.

'You have to get to that point where you literally fall out and don't talk, but then you are like, "That is out of order and from now on I won't do that, or I will do this." As long as there is a kind of change or modification in the behaviour, I think it does help to reach that point where you fall out.

'To be honest, if you are together you are supposed to have arguments. If I was in a relationship and I never had an argument at all, I would be really worried that something was going wrong. It is all part of a process, like a kind of cleansing, like a thunderstorm clears the air and you start again.'

I asked Llanzo whether he'd ever thought about ending the relationship during one of these thunderstorms, and he replied: 'Yes. I thought that yesterday. I thought about it first thing this morning … Sometimes you do it as a threat … But I thought, if I do my own thing I am going to be incredibly miserable for the first six months.

'We have been together such a long time. I am in that relationship and I like being in a relationship. I don't like to be a single person. I haven't been single since my early 20s. I am used to having a family and a partner. If I did become single I would revert to my childish days, but when you are in your 50s it's not the same. You can't be that old guy in the club with all these young ladies, being a nuisance.'

I asked Llanzo how often he had these kind of arguments, and he said: 'Once in a blue moon, every six months or a year – there is no frequency, things just build up and then something sparks it. You have to get things off your chest, don't you, because you don't want to be like a pressure cooker building up and building up, because when that blows, that's it and everyone has to leave the room. If you get into an argument every now and then, you are letting that pressure out. Some people internalise things and they will either harm themselves or they will harm somebody else.[*] It is always good to get things off your mind.'

Most of the people I spoke to agreed that arguing (if done in the right way) was generally a good thing.[†] I interviewed Henry and Mo after they walked out of the British Library holding hands, and Henry stopped to help Mo get down the steps at the front.[‡] Mo was very small with short curly hair and Henry wore a navy cravat. 'We're a bit long in the tooth. I'm 83 and Mo is 88. Adolf Hitler was responsible for us meeting because at the outbreak of war, I was evacuated from Manchester to Staffordshire, where Mo lived.

'I was eight years of age when I was evacuated and Mo was 13. I had to join up to do National Service when I reached 18 years of age, which was after the war. When I came out of the army, I met Mo at a dance. We had a few dances and then I asked if I could take her home. We were courting for just over 13 months and then we were married.

---

[*] In this chapter I'm talking mostly about conflict that is not abusive – I talk about abuse in the next chapter.

[†] Both the interviewees and the research suggest that how you argue is incredibly important – but I'll come back to that.

[‡] You met Henry and Mo in Chapter 8.

We were more sexually attracted than anything – she is a very attractive woman, as you can see.'

Mo (who didn't say much) added: 'I was, I was, I'm not now.'

'Mo and I have been married for 62 years. We've had our ups and downs, but we've never separated. If you don't agree about something, I think you have to be honest – you don't have to start fighting about it. If there's a difference of opinion, try and reach a mutual agreement. We differ from time to time, but we've supported each other.

'There's a certain mentality that likes to be told what to do. We've known couples where he's said, "We've never said a cross word, have we?" And that's true only because she has done what he wanted her to do all the time.'

Mo added: 'You've got to give and take.'

'I have a lot to say. But Mo is more concise – she's right when she says it's give and take. "For better or worse" is what we said – and that's how we've kept it.'

Speaking to Llanzo, Henry and Mo made me think once again about the role of beliefs and expectations when it comes to relationships. If you expect to have no arguments whatsoever, or that your partner will be able to read your mind without your having to communicate, you're likely to be disappointed with anyone you date.* Or if you believe that disagreements are damaging and something to be avoided, when one does come about, you might unnecessarily question your relationship or worry about it ending.

If, on the other hand, you believe that disagreements, and even arguments, are inevitable, conflicts are unlikely to feel like such a threat. Instead, you might start to see them as a chance to get to know each other, to deepen your understanding of each other and solidify your relationship.

---

* Equally, if you think it's normal to have explosive rows all the time, where what's coming out is not meaning but rage, you might fail to leave a relationship that's toxic.

This was how Maurice, a 95-year-old poet who had been married for 65 years, saw it. 'If you have these arguments, it's only another way of clearing the path for each other and getting to know each other. I didn't like criticism but, more often than not, if my wife criticised me, she turned out to be right and it was well deserved.'

And Jennie, an artist in her 50s, said: 'They're opportunities, so to think you can skate through and avoid them, you're probably missing a trick. There's something to be learnt by approaching arguments, seeing what happens, how you deal with it, how you reconcile it – and whether you grow from it.'

What are we arguing about here?

When psychologists asked almost 600 engaged and married people across the US what they argued about, they found that most disagreements were about money (in first marriages) and children (in subsequent marriages). Other popular topics included: communication, sex, household chores, jealousy, friends, careers, in-laws and hobbies. But psychoanalyst Erich Fromm would question whether this was really what the discussions were about.

Were they arguing about their partner's children because the children were actually a problem, or because they felt secondary or rejected? Were they arguing about money because they really needed more, or because they associated money with security and no amount would ever be enough? Were they arguing about chores because they didn't want to wash the dishes, or because the chores reminded them of the way their parents didn't work as a team?*

---

* This is obviously not intended to be an exhaustive list. In fact, I'm pretty sure such a list is impossible.

Fromm believes that most arguments are actually attempts to avoid the real conflicts between two people. Instead, couples argue about things to cover up how they really feel, or project their own feelings, or feelings from other relationships, onto their partner.

He believes that the real conflicts, those that are experienced at a deep and honest level, aren't destructive – they're cathartic and clarifying, and help people emerge with more knowledge and strength. But to have this kind of argument requires each person to be vulnerable. They won't achieve it if they're running away from themselves. Instead, each partner must come from the 'centre of their existence'.*

Chloe was a receptionist in her early 40s, with short, impish hair and high cheekbones. Her husband Tim died when he was 31. 'The only thing we would really argue over was our dog. Tim loved his games console. We didn't have a garden, so at 5.00 a.m., then again at lunchtime and in the evening, I would be walking the dog. I'd ask Tim to do it and he'd say, "Yeah, yeah, yeah." I'd ask again after half an hour, then half an hour more, and he'd blow up and say, "Look, I said I'd do it. I just need to finish this first." So, I went ahead and walked the dog, even though I wanted him to do it. What was the point of arguing if it wasn't changing anything?

'I guess I did most things in the relationship. It would be me doing all the housework, the cooking, the cleaning – and when I asked him to do any of those things, I'd get the same thing. He'd go to work, come home and that would be it, games console on. Then it would be me having to sort the dog out, the cooking, the cleaning. So, when we argued about walking the dog, we were probably arguing

---

* Although Fromm doesn't actually define what he means by this, I think Jennie touches on it later in the chapter. Either way, I don't think it's easy to do.

about an inequality in our relationship, with me having to do most things.

'I don't think he realised until he got sick. I think when everything is normal and tickety-boo, you don't notice the ruts you have fallen into. It wasn't until he got leukaemia that he said, "Thank you so much for everything you have done for me, I didn't really appreciate how much you did." I think, for him, being ill brought into focus how much I did and how much he didn't do. Maybe it was to do with dependency; he was completely dependent on me when he was ill. But I don't think it entered his head before then.'

Maurice also noticed the discrepancy between what his marital arguments seemed to be about and what was actually going on. Having long been a fan of his poetry, I travelled to his retirement flat for an interview. We spoke surrounded by his daughter's artwork and a photograph of his wife, Olive.

'We had good barneys. She threw a plate of bread and butter at me once. At that time, if she'd had somewhere else to go, she would have walked out on me, but we would have got back together the next week. We had good old ding-dongs.

'An argument would never drag on, and we would usually hug and make up – but never quite as quickly as that, because one of us would be hurt by assumed or real criticism. Some people say don't go to bed on it, but if you sleep on it, you very often wake up with a little bit more of the person's thinking in your mind than your own, so you're better able to come to a satisfactory conclusion.

'Behind most of our arguments was a state of mind being brought about by various factors, no one particular thing. I could not argue objectively, only subjectively. I would respond exactly how my heart says, then think after. Now that is not a very clever thing to do. I still needed the rough edge chiselling off.

'I didn't like family rows because as a child I heard my mother and father arguing. I lay in bed listening to them and I would imagine that horrible things were going on, which weren't. My father had a bad temper after drink and he was usually annoyed with himself for having let himself down, but I didn't know that until later.

'The war had just finished. Nowadays, soldiers in the army get diagnosed with PTSD and get counselling, but not back then. In the early stages, when I was being rehabilitated into civilian life, many of the arguments were because I wasn't settling in, I wasn't as relaxed as I should be. But we got over it. It was just a period of getting to know each other.'

Intimate relationships, and the disagreements and compromises that are a necessary part of them, can trigger parts of us we keep well hidden, even from ourselves. That means that a disagreement that's ostensibly about returning phone calls might actually be about someone's sense of security in the relationship and the activation of their hugely anxious attachment system. Whereas a dispute about wardrobe space might really be about a highly avoidant person's sense of suffocation and their need for independence. According to attachment theory, the underlying cause in both cases is a fear of loss.

While getting to the heart of some arguments can be relatively easy, in others it can be incredibly difficult. In fact, it can be a lifetime's work.

'When you're arguing, you're trying to find the depths,' explained Jennie. 'What are we arguing about here? What have you got to lose? Are you being criticised? Come on, where are you? What are you really about? Let's get that level of communication going ... There's so much banal stuff around, but you've got to find the heart of an argument, of someone.'

Accommodation

If someone starts screaming or sniping about something you've done wrong, it's all too easy, tempting or automatic to scream or snipe right back. But another option is to stay neutral – or maybe even try to respond in a positive way. This, the decision not to respond destructively but constructively, is what psychologist Caryl Rusbult and colleagues call 'accommodation'.*

'I bite my tongue, really quite badly,' explained Chloe. 'I just think to myself, "It's not worth making things worse." I count to 10, or if I get an angry text, I wait until I can write something calming. Sometimes I even apologise, even though I don't think it's my fault. If you start screaming and shouting at someone, right from early on, you're never going to get anywhere, it's counter-productive. It's like with your children – don't start screaming at them if you want them to do something because they're not going to respond.'

Accommodation is particularly important if you're going out with someone who has an insecure attachment style, but if you yourself are unresolved, you might struggle with it. That's because people with an unresolved attachment style can live in a world of triggers and be prone to both irrational thinking and explosive behaviour when they're very stressed. This can have a similar affect in arguments to pouring kerosene on a naked flame. You might also struggle with accommodation if you have an avoidant attachment style – highly avoidant people tend to care less about conserving intimacy (in fact, they're often subconsciously trying to avoid or destroy it), and this means they have much less motivation to overcome the impulse to argue or withdraw. Instead, they tend to get angry and indulge their desire to disengage – the theory

* Just to be clear, I'm talking specifically about the work of Rusbult and colleagues – other psychologists have used this word to refer to something very different.

being that this blocks out feelings of rejection and restores a sense of personal control.*

Ada, from Beijing, China, described how, after many years of quarrels, her partner had learnt to accommodate her tendency to get angry and start arguments.

'Since I have grown up and stopped behaving like a childish brat, basically we don't have arguments any more. We do have disagreements sometimes and I am quite hot-headed, but he knows me so well … He will wait until my hype has gone down and then suggest something different, and when I come to my senses, then I know that I didn't make sense and that he was right.

'Before, if he didn't agree with something, he would say straightaway and he would be arguing really badly. But now he knows that I am a two-second person: I will say something and then in two seconds, I have forgotten about it and I have moved on. Then we will discuss it rather than have an argument. And a lot of the time, he is right.'

Another way to take the heat out of an argument and accommodate is to use what psychologist John Gottman calls a 'repair attempt'. It's basically anything you say or do that diffuses negativity and keeps a conflict from spiralling out of control.

A few people I spoke to described ways in which they tried to do this. A nine-year-old I interviewed in Canada told me she pulled monkey faces in arguments if they were getting too heated (she then did a full-body monkey impression for about 10 seconds). There's something to be said for facial (or full-body) impressions as it can be difficult to make a verbal repair attempt, or hear someone else's. That's because, in the heat of an argument, it's easy to deliver a good message with an angry voice, and it's just as

* One way this might work is by reducing how much their partner can hurt them.

easy to hear the angry voice and not the good message. Research suggests that tone is about five times more influential on the way we interpret what someone is saying than the content.

But humour doesn't work for everyone and repair attempts have to be crafted to suit the individual. When I asked Llanzo whether he and his wife ever pulled faces to try and calm arguments down, he said, 'We have had arguments and … I have been physically attacked for pulling a face. She takes it as me laughing and goes mad. It is like I am trivialising the whole situation and taking the piss out of her. "This is not a joke," she says.'

There's no getting around it – cooling down an argument isn't easy and can take a lot of effort. But it's possible. And it might get easier with practice if the well-being of your partner and relationship become more important than the urge to retaliate, or seeing things from someone else's point of view becomes more of a habit.

Maybe that's why couples who manage to make their relationships last until old age show fewer negative emotions than middle-aged couples when they're having an argument – and have fewer arguments in the first place. They get better at prioritising each other (and their relationship) above their impulse to be mean, scream, win an argument or walk away. (Either that, or they get more placid as they age, or they were always very good at it and that's why their relationships have lasted.)

But accommodation at absolutely all costs isn't the aim either. 'It's probably not always very good for me because I'm not saying how I feel. I'm probably a bit of a doormat and let people walk all over me. But sometimes it's good because it avoids conflict,' said Chloe. Maurice added: 'You need to take on board the other person's opinion, but you need to be honest. The worst mistake I think you could make would be to pretend to agree with them just to keep the peace when actually you don't.'

Self-control (again)

To the extent 'accommodation' means resisting the urge to retaliate, psychologists Eli Finkel and W. Keith Campbell argue that self-control plays a crucial role. You'll hopefully remember Jessie, who stopped cheating because she changed the way she saw the temptation of infidelity. Instead of focusing on the sight, smell and touch of a handsome man, she looked at infidelity as an unenjoyable and even disgusting experience ('You feel like a whore afterwards, when you have to get dressed in front of them'). The temptation stayed the same – it was Jessie who had changed. The power is often not in the stimulus, but in how we evaluate it.

This strategy, of changing how you see things, might also help in resisting the temptation to yell back when someone yells at you.* Instead of seeing yelling as something your partner deserves, you could see it as damaging and hostile, something that numerous studies have linked to marital misery. And instead of seeing the person in front of you as someone who is criticising or undermining you, someone who you need to defend yourself against, or put back in line, you could see them as a flawed human being who's genuinely upset about something that's really important to them (or is snapping about something that isn't). Or you could try to zoom out and see them as someone who's had a particularly bad day, someone who's sleep deprived,† hungry, or someone whose life up until

* Yelling is only one of many ways to keep an argument raging. Carefully chosen words delivered calmly can be just as, if not more, inflammatory.

† There's some brilliant research on the impact of hunger and sleep deprivation on our thinking, feeling and behaviour – for example, after being awake for 19 hours, people were found to be as cognitively impaired as people who were drunk – so it might be an idea to avoid very late-night arguments if you or your partner got up early.

that point may have not provided them with the best ways of regulating their emotions or communicating how they feel – someone who needs your compassion, not your criticism.[*]

This was something my interviewees described in their own lives. 'We used to have a tiled floor in the kitchen, and when we had arguments, we'd sit at opposite sides, right against the wall,' explained Jennie. 'I was actually outside myself watching these two people who loved each other argue, the farce. Once I'd glimpsed it, then I'd find a way to move – "I will take one step across this tile if you make one step." Sometimes it didn't work because the timing was so bad, sometimes it was explosive. But I still stuck with it, and eventually I could make it across the floor. Not always, but mostly.

'At the time, it feels like a hugely important argument about something catastrophic. But it wasn't catastrophic, was it, because you could make it across the floor if the timing was right. I learnt that from my husband. It didn't always work, but when it did, it was so liberating. You had to be prepared to just give.'

Paul, who I met in Kashgar, China,[†] added: 'I try to get all the factual information that I can, then I try and empathise, I try to zoom out to see why I'm feeling that way, why my boyfriend might be feeling that way. I think it's about taking the bigger picture, understanding that everyone gets upset and it might not be logical, and there might not be much you can do about it. It might just be the way that they're feeling, and your role is to try and understand that and work through it.'

---

[*] Of course, there are limits to this – if someone continually unleashes their rage on you, there comes a point when it may be best to end the relationship. I'll explore this in the next two chapters.

[†] You met Paul in Chapter 4.

Research suggests that zooming out really does help. In a long-term study of married couples in the US, half of them (let's call them Group A) were asked every few months to think about one of their disagreements from the point of view of someone neutral who wanted the best for everyone involved. They were then asked to think about what stopped them from adopting this perspective during arguments, and to try to take this point of view in the future. The other half (let's call them Group B) were given no such instructions. Unsurprisingly, Finkel and his colleagues found that Group A felt they had better relationships than Group B and were less upset by arguments.

Another strategy is to focus on what you would get if you didn't give in to the temptation of shouting back. When researchers showed children pictures of what they would get if they waited, they were able to delay eating the marshmallow in front of them by an average of 18 minutes. Of course, there are no experimenters lurking in cupboards to jump out just in the nick of time and show you pictures of what you'll get if you accommodate, but perhaps you could imagine it.* You could, for example, fast-forward a few years and imagine being happily married for over 65 years like Maurice, or helping your partner on the steps of the British Library when you're in your 80s, like Henry and Mo.

Or you could imagine the long-term costs of giving in – when smokers were asked to focus on the long-term negative effects of smoking, their cravings dramatically reduced. In the context of an argument, you could picture screaming, slammed doors, smashed plates or a child in the next room worried that its parents are going to break up.

---

* Neuroscientific evidence suggests that our imagination is hugely powerful and engages areas of the brain that would be involved if you were actually experiencing what you're imagining.

Longer term, you could imagine more and more conflict, an increasingly unhappy relationship and, finally, the devastation of break-up or divorce.*

I appreciate this may all seem a little unnecessarily imaginative, but cool reasoning may be no match for the hot temptation to retaliate, so the 'hotter' and more emotive the images you can conjure up when you want to shout, scream or stomp, the better chance you might have of holding back.

The art of arguing

There's another way of looking at accommodation. Most of the research assumes it's difficult and takes a lot of effort – but that implies that what you really want to do in the heat of the moment is scream, rather than act in a way that might be better for your relationship.

When I asked Henry and Mo about their arguments, Henry mentioned the importance of give and take in the same breath in which he talked about their commitment to each other and Maurice emphasised the importance of generosity, saying, 'You have got to be prepared to go a little bit beyond halfway. Unless you are prepared to do that, when you get near halfway, you are going to be looking for the halfway mark to stop.'

Perhaps the more committed you are, the more you see yourselves as a team, the more you value your partner's well-being, and the more you practise seeing things from their point of view, the easier it might be to not only stop

---

* We explored the idea of 'bundling' decisions in Chapter 9 – if you think your present decision will impact your future decisions (retaliating now will make you more likely to retaliate in the future) you may be less likely to do it. Although if your attachment system is subconsciously driving you to be alone anyway, this strategy may not work that well.

yourself from retaliating, but also to respond in a kind, compassionate and loving way.

Maybe accommodation involves giving, which, according to Fromm, is not a limited resource. Instead of being something that depletes, deprives or involves sacrifice, he argues that, through giving, we can experience strength and joy, as well as what it is to be overflowing and alive.*

Perhaps, like love, arguing is also an art – and the skills needed to have helpful and loving arguments are the same skills that are needed to master the art of loving more generally. Surely the generosity, empathy and compassion that will help turn potential fights into constructive discussions are the very same attributes that transform the way you deal with every other human being.

By zooming out and seeing the person in front of you in the larger context, by recognising they are acting out of fear, frustration, exhaustion, upset, anger, trauma or stress, giving them the benefit of the doubt instead of assuming the worst and looking at things from your own point of view, I suspect you can transform most conversations.

* Although I'd argue that ceaselessly giving to people who abuse or take advantage of us is unlikely to be good for our mental well-being.

CHAPTER TWELVE

I Woke Up in the Gutter

'I suddenly thought, "This is not a partnership – we're on two parallel lines and they're never going to bisect. It's just not going to happen."' I met Angela on a radio production course. It was pelting down outside, so we ventured downstairs and eventually found a quiet room to speak in. It was long, narrow and sparsely decorated, with little more than a navy sofa squashed in the corner.

'I realised he had a bad drink problem, I realised he was a liar because of it, and I realised that he was not violent, but very destructive with other people. We worked together and there was another young guy who had just got his first car. It was clear that my fiancé was either jealous of it or was in one of those moods. It was the end of a work do or something, and he just literally got his car and rammed this fairly young guy's car again and again with me in the seat. And I thought, "Why have you done that? He's done nothing to you, he's just got a car ..."

'I wish at that point, I'd have thought, "I really, really hate this," but I was quite – my mother called it – "intoxicated" with him. He made her hair stand on end. She couldn't bear him. And at one point, long before it got dangerous, she must have said something.

'My older sister died when she was tiny, an illness sadly, and I remember her sitting there one night and something had happened. I was caught between her and him, and she was sitting there ironing, and I remember as if it was yesterday, "I've already lost one daughter and I don't intend to lose another."

'We had an argument as I caught him getting together with somebody else we worked with and I confronted him.

He was drunk and he chased me out of the house. I think I was better with words than him and I had wound him up, I admit I wound him up. He ran out of the house, dragged me into the street and just started rabbit punching my temples and behind my head, and I passed out.

'I don't remember anything past the first one or two, so I don't know how, but I ended up waking up in the gutter – really nice, Sunday afternoon, in a nice street – just feeling really embarrassed and feeling people watching. Nobody came out of their house – that is a laugh isn't it – but feeling people watching. And this was the scary thing, I went back in the house and the red mist descended. I picked up the bottle that he had been drinking out of and slammed it around him, and then staggered off home.

'It was a boss of mine who just knew on sight when I walked into work the next day looking a mess … She dragged me into an office and told me her history and said, "You are not going back there." It was as if my friends and work colleagues closed around me and made plans to keep me away from him … This all ended up with me getting in a car with another mate and driving to the south of France to a friend for the next week. And if I hadn't done that, I would have gone back.

'What has struck me since is that most women wouldn't get the chance to do that. They might have kids. They might have no money. I had it and I could move away and I could get away. I've never seen him from that day to this. And I feel very sad for women who haven't got that boss and those friends to pull them away and just get them away for however long it takes to take a look back in and say, "That's crap, I've got to be out of it." I can understand as the day it happened, that night, I was saying to my father, "Don't hurt him, it was my fault as much as his and I hit him as well." And the next day, I was on the phone, saying, "Please, please …"

'My boss was older than me, so I guess she was born in the late 40s. She was a virgin when she got married and was raped repeatedly on her wedding night. No questions, "You are my wife, this is what you will do." No intimacy, no tenderness, just rape and lots of violence.

'She left the next day, she was just so terrified and so horrified. I don't think she knew what she thought from what she said. Someone took her in, once again it was a colleague, and it was at least six months before she told her parents that she'd actually left. Unbelievable. But I know that happens.

'She made it very clear to me that people can die and I was like, "He only hit me once." She said, "Women die all the time from this sort of behaviour," and persuaded me that there was a life away, a better life.'

Angela's words reminded me of something I heard from Donny,* who, as a policeman in Glasgow, Scotland, came into contact with partner violence† cases every day. I asked him why people didn't leave and he said: 'Some of them stayed because they were either scared senseless or trapped financially, and some of them never left at all because they got murdered.

'I have seen it both ways and men get abused as well. They were terrified. These are the guys you would see in the pub

---

*You met Donny in Chapter 5. He met his wife on the dancefloor when he was 16 and they had been together for 50 years.

† Unhelpfully, different phrases are bandied around and can mean very different things, depending on who is using them – 'intimate partner violence' can refer to physical violence as well as other types of abuse, while 'domestic violence' or 'domestic abuse' can describe partner violence, any violence that happens within the home, or non-physical types of partner abuse. I'm going to use the phrase 'partner abuse' to refer to physical, emotional/psychological, sexual or financial abuse, or coercion or controlling behaviour (which I'll explain shortly). When I'm talking specifically about violence, I'll usually call it 'partner violence'.

with a scrape on their neck or a tuft of hair missing and things like that. I'd think, "That's a big guy, I wouldn't want to tackle him …" and his wife is a 7 stone (98lbs) woman that is beating him up. He wouldn't defend himself and she would hit him with an iron or run her nails down his face.'

It wasn't until I interviewed Angela and Donny and delved into the numbers that I learnt just how common homicide* was in the context of romantic relationships. In England and Wales, 22 per cent of all homicides between 2013 and 2016 were caused either by the victim's partner or an ex.† That means two people a week are killed by someone they are or were in a relationship with. And victims are usually women (most female homicides (52 per cent) were caused by a partner or an ex).

Breaking up doesn't guarantee safety either. In 2016, 40 per cent of partner homicides in England, Wales and Northern Ireland happened after the couple separated. The first three months were the most dangerous‡ – from then on, the risk decreased steadily over time.

Even when someone escapes death at the hands of their partner or their ex, they're more likely than those who haven't been abused to take their own life. Research by the World Health Organization (WHO) found that, in each of the 11 countries examined,§ women who had

---

* Homicide includes murder and manslaughter.

† To be specific, this percentage applies to homicides of people aged 16 and over between April 2013 and March 2016, recorded by the police in England and Wales.

‡ In the Femicide Census (which collates the details of femicides committed by men), half of post-break-up homicides in England, Wales and Northern Ireland by an ex-partner were committed in the first three months apart.

§ The countries involved in the study were: Bangladesh, Brazil, Ethiopia, Japan, Peru, Namibia, Samoa, Serbia and Montenegro, Thailand and the United Republic of Tanzania.

suffered partner violence were more likely to try to kill themselves than those who hadn't.

Most abusers don't go as far as murder. Instead they grind their partners down with incessant, and often increasing, abuse. It's this type of abuse (the non-killing type) that's difficult to get exact numbers on because there's an accepted level of privacy surrounding romantic relationships, which abusers can take advantage of. On top of that, many keep their suffering secret.* Even so, studies around the world suggest that it's uncomfortably common.

In the WHO study, the percentage of women who had suffered physical violence by a (male) partner ranged from 13 per cent in urban Japan to 61 per cent in provincial Peru. In most of the countries examined, between one in four and one in two women had been (or continued to be) physically abused by a partner.

Severe physical violence (being hit with a fist, kicked, dragged, choked, deliberately burnt, threatened with a weapon, or having a weapon used against her) was also devastatingly common, ranging from 4 per cent in urban Japan to 49 per cent in provincial Peru. And the vast majority of women experienced acts of violence more than once. This means that when someone says, 'It was just a one-off,' it's highly unlikely to be true.

Although most people who suffer partner violence are female, as Donny explained, men are also abused. Just how many depends on the study and the country, but the numbers aren't small. In a national study in South Africa in 2001, for example, 21 per cent of men (and 29 per cent of women) admitted to experiencing violence in their most recent relationship, and in a 2006 UK study, 35 per cent of

---

* In a 2006 UK study of same-sex relationships, 22 per cent of the people questioned who admitted to experiencing partner abuse (physical, emotional or sexual) did not seek help from *anyone*.

men (and 40 per cent of women) had experienced partner violence in a same-sex relationship.

What's especially interesting about the second study is that some of the people who were maltreated didn't see it as abuse. When completing a questionnaire at the start of the study, some said they hadn't been abused, but when they went on to be interviewed, they described what were abusive experiences. This is another reason estimates should be taken as a minimum – not only do some deliberately keep their suffering secret, some don't realise they're being violated.

The disappearing self

'He was very manipulative and I think this is a feature of most abusive partners,' explained Angela. 'They manage to strip away your confidence in little ways so that you don't really notice. They make you feel that they are the only one for you. I was very young – I was 22.

'Maybe after a year or 18 months, I noticed that I was losing my confidence. I could see it if I look at photos – I didn't look as good, I didn't see certain friends, certain friends were missing, which always upsets me that he managed to cut those out of my life. I just looked clingy, unsure.'

Angela wasn't the only person to tell me about the slow but steady erosion of confidence and character. Valentina, a restaurant manager in her early 30s from Spain, explained: 'I was just crying more than smiling, and very miserable. I was a very jolly, happy, friendly person that loved travelling, loved going out, meeting people, and suddenly I became like a grey cloud. I wasn't happy at all.

'When you buy a flower that is beautiful and pink and suddenly becomes a bin full of shit and broken bottles, you think, "How can that develop into this thing?" You don't expect that change at all. I don't think that anyone will buy a flower that becomes a bin. The funny thing was that I didn't want this person in the beginning. I was fighting against his flirting for a long time and suddenly one day I said, "Why not?" and I fell in love completely. And once I was in love, he relaxed and showed his real face and I became a sad, miserable person for a long time. That is the worst experience I have had with love.

'First of all, my friends said, "It's good that you have a relationship, but it's not a healthy one, it's toxic." Then they said that I became another person because he was very jealous, to the point that a gentleman that was probably 70 or 80 in the street stopped him and said, "Your partner is beautiful, you should be really proud." He was like a grandfather for us in terms of age – but he spat at him. I was terrified. "What are you doing, he was very nice to you and you have taken that as something bad!"

'Can you imagine? In the middle of the street at 12 o'clock in the afternoon, poor gentleman just says something nice and that reaction? He changed me in the way that I wore my clothes and the times that I went out at the weekends. If a guy passed when I was walking, I would always look at the floor because he would overreact so badly.

'I think that when you are in these situations where you're completely blind, whoever is telling you something that is going against your partner is your enemy – you are not going to listen, never. I always say when I am giving advice to friends – from the outside everything seems very clear and easy, but the only person that is going to realise the decision that you need to make is yourself. And sometimes it is when you are on the edge of the cliff, close to losing everything.'

Valentina's words reminded me that our intimate relationships have the potential to change us in important ways. As I explored in Chapter 4, philosophers argue that a good partner can help you become a better person and long-term research suggests that, over time, couples rub off on each other, growing more similar in happiness, intelligence and vocabulary. But if your partner is abusive, the change is very rarely for the better.*

Changes may be subtle at first, barely noticeable even (I doubt, for example, that anyone has ever thought, 'Wow, my vocabulary really is improving thanks to this relationship'), but slowly and steadily, they can have an enormous impact. And abusers rarely rely on just one technique to cause their victims to crumble.

Although violence may be the most visible form, abusers often use a number of other, less obvious and less physical techniques, including emotional,† sexual or financial‡ abuse, as well as using bullying or control. They may operate in different ways, but they have very similar, and destructive, outcomes.

Emotional abuse is an attack on someone's personality and esteem rather than their body. It can involve putting them down, making them feel like they're going mad and blaming them for the abuse, or humiliating them in front of others. It can also include bullying or control. Bullying involves humiliating, intimidating or assaulting someone to punish, hurt or scare them. It can also mean threatening – either them, or someone they care about (including their children or pets). Bullying can be a pattern of behaviour, but it can also be a one-off.

Control can involve depriving someone either of their independence, or the means to resist or escape. In one study, for example, a woman described how her partner

---

* That said, suffering can sometimes lead to positive change – psychologists call 'post-traumatic growth' and I will describe this in Chapter 13.

† Some people also refer to this as 'psychological' abuse.

‡ I've based this definition on the UK cross-government definition of domestic violence or abuse – which also includes abuse by family members as well as partners or exes.

locked up all the shoes in the house so no one would go anywhere while he was out; in another, a man had no access to money so didn't feel he was able to leave. Control can also be less practical and more psychological, for example, policing someone's everyday life (for example, Valentina's partner controlling what she wore and when she went out), or making them feel inferior or helpless by isolating them from their support networks (like Angela's fiancé slowly cutting her off from her friends).*

Sexual abuse is any sort of sexual activity that happens without active consent.† It doesn't have to involve physical contact – it might only involve words or photographs (including 'revenge porn'‡), or it might mean preventing someone from using contraception. Sexual abuse can also include very physical contact, as is the case with sexual assault or rape.

Financial abuse means controlling someone's ability to secure, use and keep their own money and resources. Abusers might stop their partner from earning or accessing their own money, they might spend or take their cash without consent, build up debts in their name, or damage their belongings (all of which would make it harder for them to leave). One person I spoke to left an abusive relationship with more than £20,000 of debts that her

---

* Another strategy is to spread rumours or lie about friends and family so they stop seeing them.

† It doesn't matter if someone consented in the past – if they don't consent in the here and now, it's non-consensual and therefore abuse. If consenting in the past was enough, that would mean anyone you'd ever slept with could force you – literally, force you – to have sex with them, and that would be okay.

‡ Revenge porn is the distribution of sexually explicit images or video without the consent of the person(s) involved, usually to blackmail them into other sex acts, coerce them into staying in the relationship or punish them for ending it. The images or video can be made with or without the knowledge of the person(s) involved.

partner had racked up in her name. She was too ashamed to tell her family or ask for help, so she paid it off herself – which took her almost five years.

Although it might be harder to recognise, research suggests that non-physical abuse is just as damaging as violence. In one study, 72 per cent of people thought that emotional abuse had a bigger impact on them than violence. I don't find that surprising – being threatened, intimidated or humiliated on a regular basis takes a vast psychological toll. In any case, the distinction between the two is rarely that clear – violence is essentially a form of emotional abuse because of the fear, humiliation, betrayal and threats that it involves. In the same study, 54 per cent of people could predict the violence they might be subjected to from the emotional abuse hurled at them.

Call in whoever you can

I asked what Angela would say to people who suspected someone they knew was in an abusive relationship. She thought for a moment before saying: 'I say this because I have a daughter who is 22 now and I know the last person she is going to listen to is her mother – it needs friends and maybe somebody who is a bit of a mentor to pitch in. And while I don't think you can ever interfere with somebody's relationship, you can try and big that person up and make them get to a level where they can see things how they really are.

'I don't think abused women can necessarily see the wood for the trees. I think someone needs to tell them there's another way. In the same way you might with somebody thinking of suicide. It will take someone, someone not close, to say, "There's another choice." But it's not always visible, and it's a case of picking your moment. It must be very tricky.

'My mother made it very, very, very clear she didn't like him, and it just caused a complete divide. So, I lost the very person who had my best interests at heart. And another typical thing from my experience is that the guy courts your friends so they can't always see the wood for the trees either. My very best friend to this day still says, "What was wrong there? What was really wrong with him? Was it your fault?" Because she was entranced as well.

'These people aren't all monsters, walking around in their wife-beater vests, clagging off the fags, getting absolutely slaughtered every night. He was a sophisticated guy.

'Call in whoever you can. If my daughter was under that threat, I think I would be talking to her friends and the other people that she trusts and whose opinions she values. I would be honest with them. I wouldn't hold back because those words can be very useful.

'I would just try to keep the communication going and keep in with her, and try not to antagonise him, because obviously that is a very easy way to get mother and father out of the picture. I'm not an expert but just talking from what I think I would do. I would try and get her into situations where she sees positive relationships as I think that can only help. But I understand that if you haven't got that support network, or if you're on your own or have a family, it's very hard.'

Angela went on to meet someone else, someone who wasn't abusive, and by the time I met her, they'd been married for more than 20 years. After all my harrowing research, I was thankful for the happy ending. But I was also painfully aware that Angela was lucky – she had the health, support network, immigration status, language skills and financial resources to help her walk away, and to do it safely.

Many don't.

He started buying me presents

'Tom had bought my vintage scooter, which had been used in a film. He wrote to me to see whether there was any memorabilia to go with it from the film. I thought, "Oh, what a charming letter." I texted and said, "Phone anytime."'

I met Melissa in Oslo, Norway. She was in her 40s, wearing a navy polka dot dress and oversized sunglasses. We walked until we found a bench to sit on and spoke under the shade of a large birch tree.

'He came round and I thought, "Oh, you're very nice." We started hanging out together, we had lots of adventures and Tom thought I was just amazing.

'I noticed that he had obsessions with things. He bought my scooter when he already had four – but that's not unusual; scooter people can err on the side of trainspotters. He had a dressing room in his house with racks and racks of shirts and suits. But the Mod thing is to wear nice clothes, so I didn't think anything of it. He was online shopping every day and I would tease him and say, "God, you're buying another shirt?" and he'd say, "Yes, but this one is different – this one has got different buttons."

'After six months of seeing each other he phoned me up to ask if I was coming down and I said, "No, I don't feel terribly well." He said, "Poor you, well stay home and I will come and see you tomorrow." He knew something was wrong. I thought, "I could lie or I could tell the truth."

'We had not used the "L word" as I don't use it when I don't mean it. I said, "Look, I don't love you and you ought to know that." He said, "Well, I love you," and it's all been downhill from there.

'For the last four months, it's been incessant presents, emails, texts, 10-page letters: "Please talk to me, please come and meet me . . . Everything I did, I did it for you, for us . . . We were meant to be together, I have given you time to think now – you were obviously upset by moving house. Once you settle down, you will realise that we were meant to be together." I receive his letters on average twice a week, the same for emails.

'It is stomach-clenching and slightly scary. Even talking about it now is slightly tummy-turny. Maybe he will set my house on fire when I'm in it.

'My responses, which have been few and far between, have been very cold, detached and factual. I have avoided emotion because that's just inflammatory. My language skills have been used to the best of my ability – talk about editing.

'After about six weeks, he told me that he hadn't been able to go to work for a month because of the break-up. His sister emailed me, asking me to be kind to him if I contacted him. I replied, "I do not talk to Tom. The best help for him is if you let him know the truth that it is over, it has been over for months and I am not changing my mind."

'A week later, I received an email from Tom saying, "I've read my sister's email – I just can't believe it. If only we could meet, you could explain why it ended and I would leave you alone." He was trying to negotiate.

'A couple of weeks ago, he sent me a letter demanding that I return everything he had given me – the letter came with an itemised list of gifts. When I moved into my house, he wanted to replace everything and had bought a kitchen tap for me. At the bottom of the list, he wrote, "Obviously the kitchen tap is fitted, so you can keep that."

'I went to the police. I wanted a restraining order or something like that, but when the police said they would arrest him, I thought, "Woah, hang on ... I don't want him arrested, he hasn't done me any physical harm. I know it's harassment, I understand that, but I need time to think about this." I wanted it reported to them and I wanted advice from them. I was and am worried that having him arrested would rile him.

'Looking back, I realise that he is obsessive about things: he changed some of those obsessions and just plopped it all on to me. Having gone through two horrendous break-ups myself, I was a bit kinder to Tom than I should have been. But now I think, "Where's your pride?" You can talk to your friends, your family, but the person you do not pester is the person who has broken up with you because you can't make somebody fall in love with you. That's the only rule in this life.'

You might think stalking is a rare occurrence, but it's more common than you might think. In a US study of more than 18,000 people, 11 per cent of women and 2 per cent of men admitted to having been stalked – and the vast majority were stalked by a current or former partner. Some research suggests that it happens even more than this.

A lot of the time, there's no obvious physical threat – just a jilted lover writing letters, sending gifts and 'bumping' into their ex in an attempt to win them back. Although each act can seem innocuous and ordinary, when they happen over and over and over and *over*, they can become suffocating and relentless. This is especially true because stalking is rarely a brief encounter. In one study, people were stalked for an average of just under five years, with 13 per cent being stalked for more than 10 years.

In case that didn't register, people were stalked for an average of half a decade. That's a long time. Slowly, but ever so surely, being stalked can erode your sense of freedom, safety and privacy – which is why it's essentially a form of emotional abuse.

A variety of studies suggest that most people – whether the general public, police officers or even those with specialist training or direct experience of stalking cases – see stalkers who are strangers as more threatening than stalkers who are ex-partners. But they're wrong: as well as being the most common type of stalker, exes also tend to be more harassing, persistent and violent. Perhaps this is a common misconception because high-profile celebrity cases (like Madonna, Jodie Foster and Sandra Bullock) spring to mind more easily than the multitude of partner stalking cases we don't hear about.*

I'm not in any way surprised that partners or exes make more malevolent stalkers than strangers. Intimate relationships tap into people's deepest, darkest insecurities, vulnerabilities and rage, unleashing emotions and trauma that they may have been completely unaware of … until they are rejected. Although, many stalkers fail to acknowledge what they're doing or why – either denying or minimising their behaviour, or blaming it on the victim ('I'm just trying to protect her', 'She deserves it', 'If only I could speak to him, he would reconsider').

What's also especially noxious about exes (or current partners) who stalk is that they usually have a detailed knowledge of the person they're stalking. They know where they like to go, who their most important people are, and what they care about most. And that means they know how to be as disruptive and damaging as possible.

There's no single profile of a stalker, although there are a few things they tend to have in common. The vast majority are men, they often have a criminal history and most are single. In one study of stalkers who were referred for psychiatric treatment, over half had never had a long-term relationship.†

* This is called the 'availability bias' in decision-making – when things are easier to remember (or more 'available'), we tend to think they happen more often. People may think celebrities get divorced more frequently than non-celebrities, when it may be just that it's easier to remember all the celebs who've divorced than those who haven't because the former are widely publicised.

† These were stalkers generally and not just current or former partners.

There's also no single, agreed definition of stalking.* Although it's illegal in a number of countries (including England, Wales, the US and Australia), definitions vary from country to country, and state to state. Even within academia, there are myriad meanings, but most researchers tend to agree that stalking involves unwanted (and usually repetitive) contact that the recipient finds intrusive or threatening.†

That contact can take many forms. In a review of 103 studies of over 68,000 stalking cases, researcher Brian Spitzberg found that Tom's approach of incessant writing and gift-giving was used by 37 per cent of stalkers.‡ Another popular strategy was getting closer to and keeping tabs on the person being stalked – which meant moving to where they lived, appearing at their work, home or in public (at their favourite shop, for example), lying in wait, monitoring or following them.

Other strategies included theft (stealing post, belongings or information, or breaking into their home); getting others to help with the stalking (like family, who might not realise what's actually going on, or private investigators, who probably do); intimidation and harassment; and coercion, physical restraint and kidnapping. Thirty-three per cent of cases involved physical violence (including harming pets, belongings or the person they're stalking).§ That's 33 per cent of stalkers generally – as I mentioned above, ex-partner

---

* Which can make it very difficult in terms of pinning down how often it happens – because different studies measure different things.

† Depending on the country/state, some legal definitions require that the reasonable person would be threatened or scared, or that the stalker made explicit threats.

‡ He was looking at stalking generally – so the numbers may be different for partner stalkers.

§ Although rates of violence varied wildly across studies, from 1–89 per cent, so it's hard to make generalisations.

stalkers are actually more likely than other types to resort to violence.

Research suggests that anyone can be stalked – there's no 'victim profile' as such. But people in abusive relationships are more at risk (both before and after break-up) than those who aren't, and the vast majority of victims are women.*

Some people are resilient and manage to cope with it,† but most pay a painful price. In a study of 18 to 74-year-old victims, people admitted to becoming fearful, nervous, suspicious, angry, paranoid, depressed, less outgoing, more aggressive and more easily frightened. Other research found that 83 per cent became incredibly anxious, 74 per cent developed chronic sleep disturbances (sometimes because they were too hyper-alert, sometimes because they had recurring nightmares, and sometimes because the stalker kept phoning them) and 24 per cent seriously considered or attempted suicide.‡

On top of psychological costs, being stalked can lead to hefty upheaval. In the study of 18 to 74-year-olds I have just mentioned, people quit their jobs, left their place of study, changed their name, moved home, changed the way they looked, and avoided seeing friends and family. Other research found that people reached out to friends, family or therapists for help, while others used meditation, exercise or drugs to cope.

---

* In Spitzberg's review, 75 per cent were. Although many of the studies that Spitzberg reviewed were of cases that had been reported to the police – so this might not give an accurate representation, say, if women are more likely to go to the police than men. That said, the gender difference seems to be a very consistent finding across studies.

† There's hardly any research into resilience in stalking cases, so I can't tell you much more than that.

‡ Because of the way the participants in this study had been recruited, the results might represent the more severe end of the stalking spectrum.

And then there's homicide. A 10-year study examining police records across the US found that, in the year before they were murdered by their partner or ex, 76 per cent of victims had been stalked. This went up to 85 per cent for those whose partners attempted (but failed) to kill them.

I wish I could provide you with some simple solutions, but there's no easy way to avoid being stalked (other than to avoid dating a stalker in the first place).* UK research found that when stalking finally came to an end, it was thanks to the victim moving house or job, the stalker starting a new relationship, finding someone else to stalk, randomly moving somewhere else, or getting sectioned,† arrested or jailed.

Although police involvement can have great results, it doesn't always – in fact, it can aggravate the situation. In his review of more than 68,000 stalking cases, Spitzberg found that 40 per cent of restraining orders were violated. That doesn't mean it's a bad idea to talk to the police, only that there's no 'one restraining order fits all' approach, and each case (and each stalker) is very individual. But speaking to the police can provide access to valuable resources and insight (for example, into the stalker's criminal history).‡

A few years after our interview, I wrote to Melissa to find out how she was and whether Tom had finally left her in peace. She wrote back to tell me that he did eventually stop

---

* Even then you may still be at risk of being stalked by a stranger, particularly if you're a public figure.

† Being 'sectioned' means being detained in hospital and treated against your wishes because your mental health puts others or yourself at risk.

‡ Although one downside of going to the police is not having your situation taken seriously, which I've read about in a few cases. Hopefully things are changing, but as I mentioned earlier, research suggests that even people with specialist training in stalking tend to see partner stalkers as less of a threat than strangers – when the evidence suggests it's the other way round.

stalking her, but only once the police arrested him, she sold the house she'd just bought and moved to a different town.

Call it by its name

Abuse happens slowly, subtly, surreptitiously, but ever so surely. And it happens a lot. It happens to adults and, in being exposed to it, it happens to their children (on top of that, kids who grow up with abuse are more likely to end up being directly abused themselves). It happens to people of all genders, sexual orientations, cultures and ages. That's not to say these factors don't impact how frequent it is: abuse is more likely to happen to females, teens and young adults; people who have a low income, low education level or are unemployed; and those who have grown up with abuse as children. But no one is too gay or straight, rich or poor, young or old to be abused.

Abusive relationships are easy to get into, particularly when, like Sue's fiancé and Valentina's boyfriend, abusers hold back on the abuse and turn on the charm. Or, in the case of a stalker, warning signs might be seen as attentive or endearing instead of controlling. But abusive relationships can be incredibly hard to get out of, particularly if the person suffering abuse has no access to money, if they're isolated from support networks, come to believe they deserve it, worry no one will believe them (particularly if their partner is charming) or that trying to leave will amplify the abuse.* On top of that, it can be very difficult to walk

* Another reason someone might stay in an abusive relationship is for 'love'. But if companionate love requires trust in your partner's goodwill towards you and their judgement of what's in your best interests, then I would argue that you can't have companionate love for someone who abuses you, because abuse is never in anyone's best interests.

away when children are involved.* (Given that sexual abuse can involve withholding contraception, having a child with someone abusive can be easier than you might think.)

Preventing abuse in the first place and helping people safely walk away once it's begun is far from easy and requires a number of approaches, including supporting those who are particularly at risk, providing the best safe housing possible for those who need it in order to leave, and protecting children who are at risk of further abuse during child custody visits.

But there's something we can do as individuals to help protect both ourselves and the other people in our lives: acknowledge that abuse exists, call it by its name, recognise how often it happens and that it's not acceptable.

In a UK study I mentioned earlier, 35 per cent of men and 40 per cent of women experienced partner violence in a same-sex relationship – but some of them didn't see it as abuse. If we don't recognise abuse for what it is, it's easier to excuse, tolerate, perpetrate and suffer. In the simple act of naming it, we recognise its existence, its gravity, its potential for harm and its unacceptability.

There's some evidence that this can create change. 'Safe Dates' is a school-based programme used in Australia, Canada, Chile, Greece, Iceland, Ireland, Japan, the Netherlands, Switzerland, Taiwan, Thailand, the UK and the US. Students take part in nine 50-minute sessions, they perform a play, and choose the best poster about abuse from a series that are posted around the school. The sessions

---

* This is why it's important to pay attention to, believe, listen to, empathise with, support and be patient with those who are suffering abuse. It's all too easy to oversimplify matters, but as Valentina explained: 'From the outside everything seems very clear and easy, but the only person that is going to realise the decision that you need to make is yourself. And sometimes it is when you are on the edge of the cliff, close to losing everything.'

focus on how people want to be treated in relationships, how abuse is defined, how often it happens and why it is difficult to leave abusive relationships. They also examine unhelpful gender stereotypes that can contribute to abuse, as well exploring how to help friends who are either being abused or who are doing the abusing.

In a study spanning four years, researchers found that students who took part in the programme experienced 56–92 per cent less abuse compared with those who didn't take part (in terms of either perpetrating or suffering abuse). Statistical analysis revealed that becoming aware of the available community support and shifting social and gender norms were big factors in this change.

If the 'they' fail to recognise abuse for what it is, fail to take it seriously, fail to believe that it is unacceptable, or fail to understand that people suffering abuse need support, it makes it easier for abuse to continue. But you and I form part of the 'they'. And that means that we, as individuals, have the power to help shift those norms, not only for the sake of those who suffer abuse at the hands of their partner, but also for their children.*

* For more information on the topics raised in this chapter, please visit lauramucha.com/resources.

CHAPTER THIRTEEN

When Things End

'When you say "love" you automatically think of romantic love, but that's not the only form. The other forms are very different. Like family love – no matter what somebody in your family does, you still love them. But it's not the same with romantic love. I was married and I fell out of love. I didn't think I was going to do that.'

Elinor and I were staying in the same hotel in Paris. We spent most mornings chatting over breakfast until her friend, who she was visiting, came to the hotel to pick her up. Elinor was in her 50s with a rasping Irish accent and short, wavy, blonde hair.

'I lived in Dublin, he lived in Belfast. When you're in a relationship, there's always a trade-off, but this time there wasn't, because speaking to him on the phone didn't affect my life. As a long-distance relationship, the lust could last longer because it was only part-time.

'We were only married for four years, and I thought, "Let me out of here," after two. When you're in love, every minute of every day you think of them, you can't get enough of them, you just can't see a time when that's going to end. And it does … Which is amazing.

'The most important thing I've learnt above love is that lust is so different to 'real' love. When you're young, you think lust, looks, body, appearance are the most important things, but as you get older, you see that it's more about upstairs, a commonality of interest* that will develop into

---

* I'm actually not sure sharing interests is vital – I think sharing values is more important. If the person you're with doesn't like football, you can just watch it with your football-loving friends – but if they don't value kindness, and you do, that's likely to be more of a problem.

love, if they have all the other personality traits that you like. When I got married, I wasn't in love, I was in lust.

'I got MS after we split up, about six weeks after.* I don't know whether the stress of that brought it on, or moving house, or moving from Belfast back to Dublin. I know you're not supposed to do those all at the same time, but I slammed them all together.

'He's now deceased. He had a massive stroke in his 50s. By then we'd been split for 10 years – I watched my own reaction, and I didn't have much of one. People kept saying to me, "Oh you must have been really sad," and I would go, "No." I must be a horrible person, because I didn't feel a bit sad.

'It would be much easier to have MS if I were in a loving relationship. Seeing friends with MS, their partners or husbands do things for them. A bit of romance would be nice in my life, but I have not been proactive in trying to get it. When you're in a wheelchair, when you're talking to somebody, they've already made assumptions about you. And they don't come near you. It's like you're on a different level … a lower level. Knowing my luck, I would get a mass murderer. Or somebody really weird. Or somebody I would fall madly in love with and he would be totally inappropriate. Again.'

As Elinor made her way back to Ireland that afternoon, I visited one of my favourite places in Paris, the Pompidou Centre.† As I sat myself down on the rooftop terrace, Elinor's words rolled around in my mind like marbles, 'When you're in love, every minute of every day you think of them, you can't get enough of them, you just can't see a time when that's going to end. And it does …'

---

* By MS, she meant multiple sclerosis. MS is a lifelong condition which can affect the brain and/or spinal cord, impacting vision, arm or leg movement, sensation, or balance, among other things.

† The Pompidou Centre is home to the National Museum of Modern Art and the sixth floor has panoramic views over Paris.

Her words brought me back to an immensely important point, the distinction between lust or romantic love and companionate love. It came up again and again and again in interviews – some talked about the fading of excitement, butterflies or being 'in love' as a reason for breaking up with someone (or being broken up with), and others, like Elinor, reflected on the mistakes they had made when high on lust.

I say 'high' on lust because that's what it is: an unreliable high. And neuroscientists now know enough about the brain to be able to explain why it's quite so powerful. When psychologists did fMRI scans of people while showing them a photograph of a person they had 'fallen in love with' a few months earlier, the brain's reward system was the most active part of their brain – whereas this wasn't the case when they were looking at a neutral person.* It's unclear whether these people were actually 'in love' (I'd argue they would only be able to know that with time), but they were almost certainly 'in lust'.

When the brain's reward system is activated, dopamine† passes from one area to another and our brain tells us to repeat whatever we just did to get that reward. It doesn't matter who or what caused the reward – it just wants more. And more. And more. And this dopamine hit has a massive impact on who we want to be with.

Humans find electrical brain stimulation in this pathway so pleasurable that some developed a strong romantic attraction to the people performing the experiments and others likened it to orgasm. For ethical reasons, human research into this has been pretty limited, but animal studies have reached similar conclusions.

---

* Of course, loving in real life is very different to looking at a photograph of someone you love. Love is not simply a response to a stimulus and scanning one area of the brain won't capture the complexity of what is going on – but it can give us an idea.

† Dopamine is essentially a messenger – a chemical that transmits signals from neurons (nerve cells in the brain) to other neurons.

Female prairie voles start to prefer one male above all others at about the same time they experience a 50 per cent dopamine increase. But when they're injected with something to restrict dopamine activity (a dopamine 'antagonist'), they prefer the vole who happens to be around at the time, even if they haven't mated with him before. Meanwhile, rats who were obsessed with pressing a lever over and over and over again in order to get the high of the reward system completely lost interest when injected with a dopamine antagonist.

When you're high on dopamine, you can't see clearly. The reward system doesn't care about who or what is giving you that high, it just wants more. You may find that, amid all the intoxication, sparkle and excitement, you manage to choose wisely. Some do. But many, very many, do not. And you won't know in that moment, when you're still high – you can only make that assessment when the dazzle dissipates. It's unclear exactly how long this takes, but based on research looking at brain activation patterns, serotonin transporter levels in blood and hormonal changes, I suspect it usually takes somewhere between 1 and 2.5 years.

Of course, there are plenty of times in life when passion may be all you want (or all you're capable of) and making commitments might be a bad idea. I spoke to plenty of people who regretted jumping into committed relationships following a break-up or bereavement, when their turbulent emotions didn't make for good decision-making.

Others lamented mistakes they'd made when they were young – and the evidence backs up the idea that making commitments in youth is more precarious. Specifically, the younger you are when you marry, the more likely to you are to divorce. In England and Wales, for example, 53 per cent of women who married when they were 20-years-old or younger were divorced within

30 years.[*] If they married when they were 30–34, the percentage dropped to 23 per cent, whereas if they were 45–49, it dropped even more, to a measly 7 per cent.

Even if all you want is a dopamine high and you're not looking to commit, it can be all too easy to slip into something without actively making a choice. The longer you're together, the more possessions, friends and bank accounts you share, the harder it can be to leave and the easier it is to stay.[†] This more passive decision-making is one of the reasons researchers believe that couples who live together before getting married are more likely to break up than those who don't (the argument being that people often have a lower threshold for living with someone compared with marrying them, but once they live together, they often end up sliding into marriage).

Either way, it seems prudent to avoid big decisions (like marrying, moving in, moving country or having children) until the drugs have worn off. But what's just as important to recognise – and even expect – is that they almost certainly will.

That's not to say lust goes completely. As Leo, from Switzerland, explained: 'Lust comes in waves. Sometimes I am horny and it is an important part of the relationship, and sometimes I'm not and then it is not. I think our love is now

---

[*] These Office for National Statistics stats compare women who married in 1976 (which allows us to see how many had relationships that lasted 30 years or more), but a number of key factors are very different now to how they were then, including women's employment rates, the increasing acceptability of divorce and the decline of religion. What was considered 'late' in terms of marriage in 1976 might well be considered 'early' today. But still, it gives us an excellent, real-life insight into the relationship between age and marriage success.

[†] All of this accidentally builds practical commitment by increasing both investment in the relationship and barriers to leaving. I explored this in Chapter 8.

more about a sense of connection, a sense of being together.' And there are things you can do to reignite passion. Recent research suggests that doing new and exciting activities with your partner, being responsive, making them feel desired and valued and spending time with other couples where you have positive, supportive and intimate conversations can all rekindle sexual desire. So too can trying to meet your partner's sexual needs without expecting an immediate return, and believing that you can be sexually satisfied in your relationship through hard work and effort.

Should you keep reading?

As I gazed at the Paris skyline clutching my notebook and pen, I reflected on an interview I'd done five years earlier with a couple in the US. They were both in relationships when they met, but, 'It was already clear to each of us that we were with the wrong people.' Pete broke up with his girlfriend, Clair got a divorce and they had been together ever since.

'My ex-husband turned out to be a jerk,' said Clair. 'But on the same note, we make the mistakes that we make, and from the experience we get a much clearer vision of what we want from another person. I don't think my values were aligned with my ex. That was the problem.'

'Clair is my best friend,' explained Pete. 'And she's a good-looking best friend, which is convenient. Plus, our values are aligned, which makes everything easy. If you realise you're in the wrong relationship, I think you have to make a clean break. If you read a book and the first few chapters are amazing, incredible, and it just kind of goes downhill from there, do you put the book down or do you keep reading the book? … It took me a while, but I slowly came to realise that it wasn't right and I decided to put the book down.'

Although there are some major differences between relationships and books (a book can't break-up with you because it has a highly avoidant attachment style, for a start), I liked Pete's analogy. I decided that, to the extent that romantic relationships are like books, the overwhelming majority have brilliant beginnings. They're intoxicating, enthralling and bursting with I-know-I-should-sleep-but-I-just-can't-put-this-down chapters.

Although some will keep you interested with their insight, empathy or wit, many, perhaps most, will become disappointing. You might find that, instead of ignoring everything else in your life to keep reading, you can wait to pick it up again. You might stop thinking about it in between reads. You might even stop caring how it ends. It might start to feel like a completely different book to the one you started with.

The pertinent question in my mind is not whether to put the book down when it goes downhill. Instead, I think the question is: are you left with an author who's prepared to be honest and vulnerable, whose judgement you trust, whose life values are aligned with yours and, most importantly, whose words you're prepared to listen to and be changed by?*

If the answer is no, it might be time to put the book down.

All you need is love

'In the beginning it is all brilliant, you have a laugh, the sex life is great and there seems to be a lot of lust and all that.' Grace was in her early 40s but looked decades younger. She was from the Philippines but had lived all over the world.

'It was almost too good to be true in the beginning, very intense, head over heels and there were a lot of promises being made. There was pressure to move in with him. I thought it was too soon, but eventually I did. He seemed to be really in love with me and I thought it was a mutual

* I'm relying on contemporary philosophers' accounts of friendship (which I think is fundamental to the success of a long-term relationship) here. I explored this in more detail in Chapter 2.

thing, but a year and a half of being together, a few months after moving in together, after I had given up a lot of my life, he decides that he has fallen out of love. His explanation was that he didn't get those butterflies in his stomach any more and he wasn't in love. Suddenly I was dropped, and at quite a significant age.

'When I moved in with him, it was very much his flat and I had to get rid of a lot of my stuff to fit in. So, he was sort of committing to me but always keeping the flat, and he knew he could afford it on his own if he had to. To him it wasn't the commitment that I thought it was or I was being led to believe.

'He didn't have much in the way of commitment to partners. He got married very young. He felt that he was in love with a woman who had a child – he left that quite early on, after one or two years maybe, and never spoke to the child again. Apart from that he had only ever had very short relationships. With me it was almost like he was ready to, and we were also of an age where you might settle down. He seemed to have got past that and was ready for a relationship, so he said.

'I blame myself because I have never found relationships very easy and I know I lack confidence; I can be quite insecure, which doesn't help. There must have been something about me that he found difficult, but his relationship history shows that he is quite selfish and wants to do his own thing. There is not much commitment there.

'I gave up a lot to move in with him and asked, "Are you sure this is the right thing?" and he said, "Yes, we are very much in love," etc. It is almost like he believed at that point that it was, and he is somebody who changes his mind and doesn't know why. So, he changed his mind and his reasoning was, "I was in love but now I haven't got the tingling feeling." But I see love as something that you work at and it grows – so I go with it and am committed to it. Real love is really knowing the other person and coping with them when things are bad … not this glamorous idea that a lot of people have maybe.'

The behaviours of Grace's ex felt very much like avoidant attachment: idealisation ('I wasn't ready to commit to anyone until I met you … we really are very much in love') accompanied by always having an escape route planned (keeping the flat, with all his stuff in it), followed by withdrawal ('I've lost that tingling feeling, I'm not in love any more'). And this got me thinking about the role of avoidance in break-ups.

According to attachment theory, people who are particularly avoidant* idealise independence, are obsessively self-sufficient and hate relying on others, so if they do make it into a relationship, they can feel overwhelmingly suffocated, often to the point where they feel they have to leave.

They can also unknowingly inflate and overestimate their positive attributes and ignore their shortcomings, instead projecting these on to their partner. And this means they can fail to notice, and therefore fail to remember, their partner's positives. Puffing themselves up, belittling their partner and failing to see or remember positive or loving aspects of the relationship means that people who are highly avoidant are adept at finding reasons to break up with people and return to their preferred state of self-sufficiency.

In idealising independence and underestimating how important their partner may have been, avoidant people can overestimate the chances of solving all their relationship problems by simply finding someone else, someone 'better'.† But the statistics suggest it doesn't always work like that:

---

* From now on, I will occasionally refer to 'avoidant people' or 'anxious people' for the sake of brevity, but I mean people with an avoidant or anxious attachment style rather than people who are anxious in the traditional sense of the word.

† Perhaps that's why one study found that 55 per cent of women who applied for a marriage licence where they or their partner had been married before had an avoidant attachment style. This is more than double what you would expect in the general public; in an analysis of more than 10,500 adults, only 23 per cent were avoidant.

remarriages in general are more likely to end in divorce than first marriages.*

Perhaps that's in part because it's all too easy for highly avoidant people to imagine that the problem is with the person they're dating, instead of recognising that they play a part. As Alexandru from Romania explained: 'You're not available anyway, the way that you relate to other people is broken. So, you could meet Beyoncé and still find faults.'

Avoidant people can also underestimate quite how miserable the process of breaking up can be, particularly as it's often more miserable for those with an insecure attachment style. While secure people are more likely to be buffered by both friends and family, as well as a strong sense of self, insecure people can lack both.

In one study, researchers approached people in the waiting room of a divorce court in Tel Aviv, Israel. Although avoidant people were very positive about their own coping abilities, they also saw divorce as very threatening and had similar levels of distress to people with anxious attachment styles. Secure people, on the other hand, felt more capable of coping and saw divorce as less scary. The researchers concluded that, while avoidant people might be able to cope with minor hassles, when attachment bonds break down, their façade can collapse and reveal their underlying insecurity.

While people with a particularly avoidant attachment style might minimise the importance of love and idealise independence, that's not what they actually believe – study after study has found that they're just very good at repressing their needs when it comes to love. From adults in divorce courts to children who have been left with a stranger in an unfamiliar place, those with an avoidant attachment style can appear calm and collected, but their

* Unless the first marriage ended because of death not divorce.

bodies tell a different story. Avoidant people do crave connection, specifically, a safe and secure relationship – according to attachment theory, that's what every human wants and needs. Yet those who are highly avoidant choose instead to live on a restrictive and punishing diet of self-sufficiency, idealising people they're not with and finding negatives in the people they do date.

Going out with someone who thinks they're superior, finds relationships claustrophobic and doesn't want to spend time with you may well make you feel less secure – even if their behaviour is subconscious and has nothing to do with you. But avoidance is likely to be even more difficult to cope with if you have a highly anxious attachment style (and anxious attachment is likely to be even more difficult if you're highly avoidant). Yet anxious–avoidant pairings, which I suspect Grace's relationship was, are relatively common.

They can also be long-lasting, in part because anxious people can be very slow to leave relationships, even if they're unhappy. And anxious–avoidant pairings often are, because while avoidant people crave space and independence, anxious people crave closeness – and the two are usually incompatible. On top of that, much of our attachment learning happens before we develop language skills or conscious memory. So, people in anxious–avoidant relationships probably won't be able to explain why they act the way they do, and will simply find themselves driven by an invisible – and hugely powerful – force. For those with an avoidant attachment style, this might manifest itself in discomfort, stress or irritation when their partner wants to depend on them or be emotionally close – which is likely to be all the more pronounced if their partner is highly anxious.

'I know my faults and I know that I am insecure,' explained Grace. 'But the relationship could have been fabulous and he would have ended it anyway. That is not to say that my behaviour might not mess up another relationship but it is not the reason why that ended.

'When I'm in relationships I tend to get more needy, more insecure, which doesn't help. My parents are very difficult and I have had quite a destructive upbringing with them, so I have got my issues there. I had a difficult relationship with my mum who was quite evil a lot of the time. She went out to work at a young age and has always been a bit of a mini adult and had to work to put her brothers through college. She is a workaholic, very strict, very controlling and shows very little love. It's not about hugging or being loving or nurturing or any sentiment, it's very cold and very hard.

'I must have given up a lot of myself and when my relationship ended, I felt like I didn't have any dreams any more, I couldn't remember what they were. I'm starting to do those things again, things that I enjoy that were maybe a bit belittled in our relationship and put to one side. In the past all my focus has often been on that person and I totally give myself 100 per cent, and then I feel a bit hard done by that they don't do the same.

'I am trying to build up my life, doing more on my own so that I have got more of a secure foundation should I get involved with someone else. I know that if I am still at an insecure point, I might find that I depend on that person a bit too much or will get hurt. I am just trying to work on myself a bit more so that I am in a stronger position to not be too needy or to want all their time.'

In many ways, people with a highly anxious attachment style are the opposite of those who are more avoidant. Although both strategies are born from a fear of loss, instead of disconnecting with their emotions or anything that might make them need others, anxious people magnify them. They worry about whether their partner cares about them, they prefer to be as close to them as possible and are often viewed by others as 'needy' or 'clingy'. They can also exhibit what psychologists call 'protest behaviour', which usually means getting angry or frustrated and not wanting to be close to their partner.

People with a highly anxious attachment style often exaggerate their needs, expressing them in a forceful and forthright way. Although the main reason for this is to get their partner's attention, the hullabaloo they create may also serve a larger purpose. It distracts them from the much deeper problem, that they see themselves as unworthy of anyone's care.

One of the reasons they can get so stressed when they're apart from their partners is that they haven't learnt to soothe themselves or regulate their thoughts or emotions. In one study, people's brains were scanned as they were asked to think about, then stop thinking about, different relationship scenarios. Researchers found that, in people with an anxious attachment style, the areas of the brain related to emotion were more activated, and the areas related to regulating emotions less so. Not only were they more prone to react to thoughts about loss, they were less likely to regulate their emotions once they'd reacted.*

When they don't feel like their partners are available, instead of thinking that they might be busy or indisposed, they often see this as a sign of their own deficiencies, helplessness and unlovability. And it's this lack of self-worth that means they would rather be with someone – anyone, even if that person is unreliable, unresponsive or abusive – than be alone.

This means that people who are particularly anxious can be quick to get into inappropriate relationships (because they dislike being alone) and slow to get out. In a study that followed 172 newlywed couples over the first four years of marriage, researchers found that the people with the highest

* While scans provide a useful insight into the brain, they have their limits, and don't necessarily capture the full and astonishing complexity of brain function.

level of anxious attachment in the first six months were the most likely to stay in unhappy marriages.*

One way of trying to escape this cycle is to 'earn security' (something I discussed in Chapter 3). It's hard work, takes a lot of time and motivation, and probably also some therapy (preferably with a therapist who has a secure attachment style). But, according to attachment researchers, there is another option – they believe that getting together with someone who is secure can improve your own attachment insecurity.

The idea is that if your partner makes you feel safe and secure, your attachment system won't play up quite as much. This means that you will slowly, sometimes agonisingly slowly, adjust your internal models and, over time, their security will rub off on you. But that strategy relies on you choosing someone secure in the first place (not always easy if you're very anxious) and staying in the relationship long enough for their security to rub off – and allowing it to (not always easy if you're very avoidant).

Attachment insecurity isn't only a problem for the insecure. I spoke to a number of people who, despite being secure, found themselves on the receiving end of insecurity – to the point where it destroyed the relationship. Given that around 42 per cent of people across cultures are insecure, even if you yourself are secure, it's pretty likely that you'll find yourself dating someone who's not.†

'There is a huge distinction between love and what makes a successful long-term relationship. In fact, I don't think love

---

* This study didn't examine attachment styles before marriage, but given the stability of attachment found in other studies, and the fact that the researchers assessed attachment in the first six months, I suspect they didn't change much, if at all.

† Based on an analysis of 10,500 adults, 58 per cent were found to be secure, 23 per cent avoidant and 19 per cent anxious. Eighteen per cent were actually classified as unresolved, but given a secondary classification.

has anything to do with it.' I met Seth in a cafe in Shanghai's French Concession. He had moved there with his girlfriend three months earlier, but they had just broken up.

'You can love people, but if you're not someone who is willing to compromise, you'll never have a successful long-term relationship because they require each person to give up a big part of their lives. You have to be the sort of person who actually wants to have that in your life. You yourself can be someone who is ready for a long-term relationship, but you might love someone who is destructive, who is not good for you or who is a commitment-phobe. Getting to the point where you recognise that to be the case is very hard.

'Looking back, I always knew that Lydia was a commitment-phobe. Her parents had a difficult relationship. We met in New Zealand and dated for a year before she asked me to move back to Shanghai with her. So, I did. Shortly after we moved, she had another freak out and we broke up.

'I'm just glad I realised quickly. We could easily have had kids and stayed here for a few years, and then I would have been in the position where she turned around to me and said, "I really did love you when we married but in the last five years, the kids have been annoying, we have no money, now it's too stressful and I've found this other guy who is just amazing, we have great sex and he thinks I am attractive." Then I would be stuck out here without any of my support network. At least if that happens when you are living in your home country, there are lots of other people in your life that can support you.

'Her behaviour doesn't mean that she didn't love me. I really think that she did. We got on very well, our life views on things were amazing and I'm sure I made her as happy as she made me, but I don't think that love really has any bearing on whether you have a successful long-term marriage.

'In 20 years' time, no matter who you marry or how you feel about someone, you'll go through periods where you hate each other and feel unloved. You've got to be a quite secure person to be able to deal with that. If you're insecure, as I think commitment-phobes are, you'll run away. Not many people are secure enough to think, "I'm going to forego any form of temptation because I don't want to disrupt my family and I don't want to harm my kids." It may have no reflection on how you feel about that other person – but instead be a reflection of your own insecurity.'

Conditions of love

I began almost every interview in the same way – by asking people how they would describe love. People from Turkey to Bolivia, Somalia to Sweden, told me that, first and foremost, romantic love was unconditional.

Elena, a hairdresser in her 30s, said: 'I'm sure my Bolivian upbringing affects my values in relation to love and relationships, but I believe that when you marry someone, you commit to them for life. If they get sick, if you meet other people, it doesn't matter. Overall, I believe that the foundation of love is to not ask for anything – love must be unconditional.'

And Neshbet from Turkey said: 'Women forgive men if they make a mistake, but not the other way round: men won't forgive women if they make a mistake. But it's not love. Love has to be unconditional. "I love you but you're only going to do what I want you to do" – that is not love. Love means you accept the person as they are.'

Only a few people I spoke to talked about the conditional nature of love.* The first was Orpheus.†

'I had a big relationship in my 20s. We loved each other but, over time, the sex inevitably suffered. We got to the

---

* Well, a few more talked about conditions after I challenged them on it, but I'll come back to that.

† You met Orpheus in Chapter 9.

stage where we were thinking about "spicing up our sex life" and I thought, "I'm in my 20s, surely I'm too young for this?" In the end, I wanted to sleep with other people.

'I went online in the fledgling days of internet dating – it was my way of sticking my toe into the water just to see what else was out there – and I got caught. My boyfriend went nuts. The rot set in and got progressively worse. I became accountable to him, but the irony was that I wasn't cheating. The interesting thing is that, in the end, I did cheat on him because I thought, "Well, I'm getting blamed for doing it anyway." So, his actions drove me to do what he secretly feared all along. We broke up shortly afterwards.

'It made me realise that you should never be jealous because it will only make the other person find you less attractive. Relationships are a contract and you have to behave well. You can't expect unconditional love irrespective of your behaviour.'

Junior was the only other person who told me that romantic love had its conditions. He was in his early 40s, had thick, floppy dark hair and broad shoulders. 'South Africa tends to be a bit of a religious country, so we take marriage seriously and love seriously … Getting a divorce is a very big thing and not something that happens very often.

'I believe that love is not just the attraction but also being able to deal with the rough times and still be together afterwards. Sticking with that person through the hard times and being able to say, "I am not running in that direction" … Knowing that person will be there for you no matter what. That's what I think love is really.'

I asked Junior whether this meant that he thought love was unconditional, and he shook his head. 'No, I don't think it's unconditional. If my wife cheated on me, I couldn't be with her. I do think there is a certain degree of conditionality. I think that my wife would have to kill someone or cheat on me, but other than that, there is not much that would make me leave her. It would have to be drastic for me to see my wife in different eyes.'

Seeing romantic or companionate love as unconditional is comforting – no matter what you do, or how you think of or treat others, you will be loved. In an ideal world, this is the love every child would receive from their parents or guardians – a sense of safety, acceptance and belonging. But romantic and companionate love are different, as is commitment – and love and commitment are not the same thing. If Junior's wife *did* kill someone or cheat on him,* he may continue to love her, but nonetheless decide to end the relationship. Or he might decide to stay in the marriage, despite the obliteration of any love he'd ever had for her.

As I explored in Chapter 8, I don't think unconditional commitment, commitment for commitment's sake, is the aim. If it was, it would mean that people in deeply dysfunctional relationships ought to continue to give love without asking for anything back, that they should accept their partner as they are, and ignore the consequences for themselves (and their children).

In close relationships, you allow yourself to be changed by the other person. If they change for the worse,† if their values change, their idea of what's important or meaningful in life, then those changes can have a big impact on who you become.‡ In many cases, these changes will be positive, but in some, as is the case with abuse, they can be desperately destructive.

---

* Interesting that infidelity was placed in the same category as murder … the evidence suggests that Junior is not alone in seeing it like this.
† Of course, they may not actually be changing for the worse – you may simply have been deluded about their goodness to begin with.
‡ There's also some evidence to back this up – I discussed long-term studies suggesting that couples grow in similarity over time in Chapter 4.

And this is why love and commitment *should* be conditional* – that's not to say I would advise walking away when you first discover that your partner eats with their mouth open or refuses to close cupboard doors. But there should be some very carefully selected deal-breakers.

Deal-breakers

Some conditions are easy to articulate, measure and enforce – someone crosses that line (murdering your family, for example) and it's an immediate strike, or at the very least, a seismic shift in the relationship. Others are harder to gauge – someone's behaviours can build over time, creeping up on you like wrinkles or grey hairs and, before you realise, your relationship and self-esteem are a whisper of their former selves.

When I pushed people on this, when I argued that love *was* conditional, and asked what those conditions were, they slowly and reluctantly listed must-haves like respect and kindness.

'If he had just been kind then almost everything else would have been okay,' said Cari from California. 'I don't think physical attraction is a condition because that may well go, but if someone continually withdraws kindness, then that's it.

'Where you are both mutually entering into an adult relationship, there are certain basic rules that you stick to, and treating each other well is one of them. If you do behave in an unkind way towards each other, you acknowledge it and sort it out. It seems inconceivable now that I was with someone who wasn't kind. I value kindness over almost everything else. And he admitted once we'd broken up that he'd been deliberately unkind.'

* Aristotle agreed. He thought a condition of love was goodness. He talked about friendship love, but the type of friendship he described sounds a lot like what an intimate relationship could or should be.

A couple of other people, when pushed, mentioned honesty, vulnerability, caring, understanding, commitment, support, self-respect, self-reflection and an interest in self-development. When I reflected on these, they struck me as what you might hope for in a best friend – which isn't that surprising given that most psychologists think it's companionate love that makes relationships last.

Of course, there has to be some leeway – there are days when pain, exhaustion or suffering can cause even the best people to fail the people they love. Humans are inescapably imperfect. And this means relationships require compassion, forgiveness and tolerance. But there comes a point when this can be dangerous.*

When no more can be done

'Some people would say no matter how a person changes, you should continue to devote yourself to that person as much as you once did. I think this could be extremely self-destructive.' It was a cold February afternoon when I spoke to philosopher Richard Kraut in Chicago, in the US. It was the only interview I had done over the phone, so it was strange not to be face-to-face. But it didn't seem to matter too much – I still got a strong sense of his warmth and wisdom.

'If the point being made is that you should not drop people quickly, that you should make special efforts to not abandon a friendship, that I think is absolutely true. Loyalty is a very important feature of good relationships and that includes a willingness to not abandon the other person when things are bad for them, or to say, "You're taking too much of my time." That doesn't quite go as far as saying it's unconditional, but the conditions should not be too lax.

---

* Often when your partner isn't meeting your needs – and isn't interested in doing so.

You should be willing to put up with a lot of discomfort and hard work, and a certain amount of self-sacrifice out of loyalty to a friend.

'But I would add there is a limit to that, especially if there comes a point where you simply cannot do any more good for a friend and putting effort into this is harming yourself. At that point, loyalty no longer is a virtue.'

I stayed for the children

I met Angelie in an empty hospital waiting room stuffed with tatty leather chairs and flickering tungsten lighting. She agreed to speak to me until the doctor called her. Luckily, he was running late. She was in her 50s and very apprehensive – she was one of the handful of people who asked that I take notes rather than record the conversation. 'It's easy to marry for the wrong reasons. When you're in your 20s, you don't think that far ahead. You don't know what you want at that age. Circumstances can influence you – friends getting married may cause you to think, "Oh maybe I'll get married then."

'In my case, it wasn't right from the beginning and I knew it wasn't right. I was too frightened to back out last minute on my wedding day as I felt we had come too far. I was very young. I had no teacher figure in my life, no parenting as such. I had just emigrated, so I had no friends. There was no one to point out to me that I should really think about this.

'It was me that left. He was really quite mean to me. There was no communication and he wasn't interested in trying. There was no love and you cannot change the way someone else feels. Eventually, I was pushed over the edge. I couldn't try any more. There was no love. No kindness.

'Mistakes are made in life and this was a big one. The children were small, so I stayed in the marriage for very many years until I became very seriously ill and nearly died. I believe that emotional difficulty eventually turns into physical difficulty.*

'They were adults when we divorced. While I think it was good that we didn't divorce when they were young, I am sure it affects children when their parents are not in a happy relationship. Children do not want to see their parents unhappy.'

As I hurriedly scribbled down Angelie's thoughts, the doctor walked through the double doors and called her in. She stood up, thanked me for my time and followed him down the corridor, leaving me alone in the dreary waiting room. As I sat there surrounded by dog-eared medical posters, I reflected on the horribly important yet daunting question that Angelie's interview had raised: whether you should stay in an unhappy relationship for the sake of the children.

When I turned to the evidence, I realised that the answer was far more complex than I could have imagined. You might think that following families for many years before and after their break-up would be a good place to start, but that raises the impossible question of when the breakdown of a relationship begins.† In some cases, relationships can start to fall apart years, or even decades, before they actually end.

Cari, from the US, told me that hers began to break down after the birth of her third child, 11 years before she realised she needed to walk away.

---

* Angelie was right: remember Lisa who suffered stress cardiomyopathy, or 'broken heart syndrome' when she discovered that her husband had been having a three-year affair? 'Broken heart syndrome' is a recognised medical condition that can happen to people who suffer massive emotional stress and is just one of many examples of quite how closely our emotional well-being is linked to our physical health.

† When it finishes is also an important but complex question – when children are involved and both parents try to and find a way to continue parenting them, their connection doesn't end when they break up. The same can be true for abusive relationships of those where one partner stalks the other.

'He pretty much stopped being kind by the time our third baby was born. Occasionally he would be nice but then it would stop again. He really was very unkind to me and I think it is partly a testament to the overwhelming love of my children that I thought, "It's okay, I can live with this."

'I had a happy life for 11 years where things were bad in my marriage because I was able to compartmentalise it. I'd think, "I spend most of my time with my children and that is an enormously brilliant part of my life, I have lovely friends and interesting work." I would have been rating most aspects of my life as eight or nine out of ten, and at the same time I might have rated my marriage as a two.

'For a long time, I thought that he could change and get better because it had been so good before. He didn't.'

Measuring the breakdown of a relationship is one thing; assessing its impact on children is another. Studies have examined school grades, behaviour, depression and anxiety, among other things, but the impact can be far more subtle. In a study of 12 to 27-year-olds whose parents had divorced more than a decade earlier, psychologists found that those whose parents had divorced were no more depressed or anxious than those whose parents hadn't.* They were, however, more likely to describe their childhoods as harder than most, wish they had spent more time with their fathers, feel that their fathers didn't love them and worry about having both their parents together in the same place for important events.

---

* Most of the research focuses on divorce, which isn't ideal as other forms of relationship, including living together, are increasingly popular and tend to break down more easily. While there are likely to be similarities across different types of relationship, there are also some key differences – for example, marrying involves a conscious decision to commit, as well as greater barriers to leaving (including stigma, admin and legal costs). In other words, it may be more of a 'practical commitment'.

As well as being subtle, research suggests that the impact of divorce can grow over time, showing itself far more in adulthood than in childhood. This means that measuring test scores one month after a break-up may only provide a very small part of a horribly complex picture. Instead, the sort of research we really need to explore the question follows parents and children for decades, measuring as much as possible.* But even then, we can't be sure that divorce is the cause of what we're looking at because there are so many other factors at play, factors that we may not even know about, let alone measure.

Bad news

Despite all of this complexity, the general consensus among researchers is that divorce is bad news for kids (although they don't always agree on just how bad it is). The evidence is pretty consistent: compared with kids whose parents stay together, the children of divorced parents are more likely to earn less, have poorer physical health, have more emotional and behavioural problems, feel less close to their parents (especially fathers), and achieve less. That doesn't mean that *all* children of divorce suffer in this way – it's far more nuanced than that. And even among those who do suffer, the differences between them and kids whose parents do stay together aren't always that big.† But they're are usually there.

The US National Education Longitudinal Study assessed an impressive selection of 13 to 14-year-old students across all American public and private middle schools.‡ Students,

* That's because studies that don't take into account multiple factors (for example, growing up in poverty as a result of the break-up) might make divorce seem worse than it is.

† Although that's not to say they're small. They're what psychologists call 'moderate' – somewhere between trivial and terrible.

‡ This means the sample is representative of all 13 to 14-year-old students in the US in 1988.

parents, teachers and school administrators were questioned at four points between 1988 and 1992, resulting in huge amounts of information about teens from about three years before to three years after their parents' divorce.

Teenagers were assessed in science, maths, reading and social studies, and their educational aspiration, self-esteem and sense of control were also examined. Compared with teens whose parents didn't divorce, those whose parents *did* had poorer results on almost all these measures.* Their test scores progressively worsened over time and their academic aspirations, self-esteem and sense of control dipped massively before divorce, before eventually improving (although they never caught up with the scores of those whose parents stayed together).†

Aside from academic achievement and emotional well-being, studies from Belgium, the Czech Republic, Estonia, Finland, Germany, Greece, Hungary, Italy, Latvia, Lithuania, the Netherlands, Sweden, Switzerland and the UK suggest that children whose parents break up have a higher risk of divorcing themselves. This is what psychologists call the 'intergenerational transmission of divorce'.

It's unclear exactly how divorce 'transmits'. Some argue that watching your parents divorce can mean that you form weaker commitments because you have seen first-hand that divorce is an option.‡ Others argue that, without a positive model to observe, you fail to learn the skills to

---

* Specifically, their test scores were lower at all times, as was their educational aspiration and sense of control. Their self-esteem was lower at three of the four points in the time it was measured.

† As with any research, it's not without its limitations. The study failed to assess parental conflict at the outset, so we have no idea how much the parents fought before they broke up. That's not ideal because research suggests that high levels of conflict are bad news for kids.

‡ And as we explored in Chapter 8, your beliefs about commitment can have a huge impact on the success of the commitment you make. And what you experience growing up can influence these beliefs.

maintain a long-term relationship. Others believe that growing up with a single parent can leave kids at a huge disadvantage, in part because this often means being poorer and that makes divorce more likely (although remarriage isn't necessarily the answer, there's evidence that multiple transitions are bad for kids too).

Perhaps this intergenerational transmission of divorce has something to do with the impact of parental divorce on children's attachment styles. A long-term study that followed (admittedly not many) children from birth, found that 18-year-olds who had experienced the divorce of their parents were more likely to be insecure, regardless of their attachment style when they were one-year-old. And a nationally representative study in the US found that parental separation or divorce was linked to attachment insecurity (specifically anxious attachment) in children.*

As well as being more likely to divorce, children whose parents break up have a higher risk of suffering emotional problems. The National Child Development Study captured 98 per cent of births in Great Britain in one week in 1958 and followed up with the children's parents and teachers when they were seven, 11 and 16, and the grown children themselves when they were 23 and 33.† The study found that children whose parents divorced had lower emotional well-being than those who didn't and, as they got older, this difference only increased (at least, until these children were interviewed at 33).

---

* I should add that the quality of the parenting (as well as the attachment style of the parent the kids go on to live with) can play a massive protective role in this, but I will come back to that.

† As with most long-term studies, people dropped out over time, and given that in this study, we're talking many years, not months, there was a significant dropout rate. But the researchers argue that this didn't impact the results (because certain measures – such as how many kids had fathers who were manual workers – remained similar despite the dropouts).

Finally, divorce can have a major impact on the parents themselves, which in turn affects their children. Research suggests that, compared with people who are married, those who are divorced are less happy, have more symptoms of depression, are more socially isolated and have more health problems.

The better option

Despite the depressing list of downsides and the significant upheaval, sometimes divorce is the best, or the least damaging, option, particularly when the alternative is growing up with parents who can't stand each other. This was the view of Miriam, who was in her 40s and lived in a small town in the south of Germany.

'They had a lot of fights and my brother and I said that they should get a divorce, but they were never together long enough in one place. My father was in the military and my mother moved one time with my father and then stayed in Düsseldorf. So they weren't really living together.

'They were always fighting and our Christmases were terrible, a legend to all of our friends because they usually exploded. To this day I can't stand people who whistle because my mother always stood in the kitchen whistling when she was having a fight with my father. For me there is nothing more annoying than whistling.

'My not having children definitely has something to do with my strange upbringing. My brother had children, which took the pressure off me. He's divorced now …'

She trailed off before explaining that she had also recently broken up with her long-term partner. 'There is a reason why I always pick weird guys. I think about that too. My mother and I found diaries from when I was 13 or 14. They were having this mean fight, they knew exactly what each other's triggers were. My father was always putting her down because she had a lower education.

I found in my diary, "No man will ever speak to me like that, never."

'If you look behind the façade, a lot of people aren't happy, really. I hope that I find the way to make myself happy. I am not sure if this is a man, a house full of beautiful things, an opportunity to watch movies or an endless supply of food, because I love food. Find out what makes me happy and do that. Content, also, as that is different.

'There are so many people who are together but not really together and this is my nightmare, honestly. I would rather be alone than be together with somebody that I can't stand.'

I'm not suggesting that all conflict is bad. Every relationship involves disagreement, and exposure to some types may help children develop realistic expectations of relationships as well as constructive problem-solving and coping techniques. But the evidence suggests that high levels of parental conflict can be very detrimental to a child's well-being.

Anthropologist Mark Flinn studied children and their families in Bwa Mawego, a rural village located on the east coast of Dominica. He did this by observing them, doing interviews and getting them to fill out questionnaires, as well as walking from house to house twice a day to take samples of saliva from the children in order to measure their levels of cortisol, the so called 'stress hormone'.* Amazingly, he did this for more than 95 per cent of the kids aged two months to 18 years for a few months every year or so for a four-year

* Cortisol plays a role in keeping inflammation down, regulating blood pressure, increasing blood sugar, and how your body uses carbohydrates, fats and proteins, as well as sleep, stress, fear and pain. Constant stress can disrupt your body's most important functions and can also lead to myriad health problems, including anxiety and depression, heart disease, headaches, memory and concentration problems, digestion difficulties, insomnia and weight gain.

period. In doing so, Flinn found that spikes in children's cortisol levels (and I mean a jump from 100 to 2,000 per cent) were usually down to family conflict or change.

In another study, actors pretended to have an angry fight in front of two-year-olds. Not only were most of the toddlers visibly upset, but they then went on to be more aggressive towards their playmate afterwards. When the experiment was repeated a month later with the same children, they were even more upset and aggressive than they had been the first time. Studies investigating the impact of conflict on adults has also reached interesting conclusions: when researchers exposed people who'd been experiencing relationship conflict for a month or more to an 'infectious agent', they were at a particularly high risk of getting ill.

Conflict is one thing, but when it becomes abuse, children suffer even more. Not only are children who grow up in a violent home more likely to be abused as children than those who grow up in homes without violence, but they can suffer some of the same behavioural and psychological problems seen in children who are abused themselves.

Given the impact of high levels of conflict on children and adults, is walking away ever the better option? The evidence suggests that breaking up in these circumstances can sometimes improve children's well-being, or at the very least, it doesn't seem to make it any worse.

In a large Canadian study, researchers followed four to seven-year-olds who were representative of all children that age in Canada at the time.* They first questioned them in 1994 and then again in 1998, and compared those whose parents divorced in that time to those whose parents did

* The study was actually much larger than this, with 16,903 children aged 0–11 taking part initially. But in order to limit other factors (including age at divorce), psychologist Lisa Strohschein decided to look at children aged between four and seven when the study began.

not. They found that the more dysfunctional the family before divorce, the bigger the improvement in children's antisocial behaviour afterwards. In other words, divorce actually improved children's behaviour when it meant ending a highly dysfunctional marriage.*

The grass is greener

Not all relationships that end are dysfunctional. Some are moderately happy, and for the children of these relationships, divorce may not provide an escape from conflict and misery, but an unexpected and devastating loss.

The National Survey of Families and Households is a large US study in which 4,460 couples were questioned in 1987–88 and then again in 1992–94.† When sociologists Paul Amato and Bryndl Hohmann-Marriott analysed the 11 per cent who separated or divorced in that time, they discovered that they fell into one of two groups. Around half of the couples who divorced did so because they were utterly miserable, in extremely high conflict and often abusive relationships. These couples were happier once their relationships ended. The other half, however, didn't argue much, weren't abusive and were relatively happy. So why did they break up?

The relatively happy couples had a number of factors that have been linked to divorce (risk factors that the unhappy group also shared, but that weren't present in the couples who stayed married). These included: getting

---

* There was, however, no improvement in anxious or depressive symptoms. But other studies have found an improvement, so perhaps it takes longer to show itself. The results also suggest that the family dynamics that make divorce more likely also increase the mental health problems of children, long before the actual break-up.

† The people questioned were representative of the US population. They were also questioned again a few years later, but that data wasn't analysed as part of this particular study.

married at an early age, living with other people before marrying, having divorced parents, having stepchildren in the household, holding liberal family values and believing in the acceptability of divorce.* Amato and Hohmann-Marriott argued that these risk factors led to the relatively happy group having lower levels of commitment.

Not only did they have fewer barriers to leaving (because they saw divorce as an acceptable choice), but they also seemed to have more 'options' in terms of alternative partners. In about 75 per cent of the relatively happy group, one or both partners were involved with someone else when the marriage ended.

Of course, it's difficult to say whether lower commitment played a role in the failure of their marriage, or whether the failure of their marriage led to lower commitment. But the fact that they had average scores on most relationship measures would suggest that their commitment was to blame, at least in part. After all, it's precisely when relationships are not particularly happy that relationship commitment comes into sharp relief.†

What I find heartbreaking about this is that, not only were the parents themselves less happy after divorce (or at least four to seven years afterwards, when they were interviewed), but this was also probably the case for their children.

In a 17-year study following more than 2,000 spouses, leaving a high-conflict relationship was linked to a relatively high level of well-being in the children.‡ But when people chose to leave a low-conflict relationship the opposite was true: their children's well-being was low.

---

* Other recognised risk factors include: being unemployed, having a low level of education, living with your partner before marriage and having a child before getting married.

† I explored this in more detail in Chapter 8.

‡ Well, to be precise, the adults were interviewed five times, the children were only interviewed twice, in 1992 and 1997.

Minimising damage

'I came to realise that he was very good at being in love, but not very good at loving,' said Cari. 'He was good at the initial passionate part, but not at sustaining love through tricky times. I lived in hope that things would improve, but I had also witnessed his parents' relationship before they split up, so I could understand why he behaved the way he did. He recognised that he was repeating patterns from his childhood.

'I remember being told what a terrible thing we were doing and that we were damaging our children. But I think the damage happened when they spent 11 years with unspoken conflict, growing up in a unloving relationship where one person didn't respect the other. When I suddenly realised that was happening, I instantly got help. My husband had grown up watching two parents where one was unkind to the other and I didn't want that to happen to my children.

'We were in counselling the next week and it was immediately very clear that we needed to separate. I would have stayed with him forever if I hadn't decided it was actually much worse for the children that way. I realised that if we'd had counselling before, we would have come up with such stark differences in what we thought was important that we wouldn't have been able to get married and would never have had children.

'When we split up, we made a commitment to our children that we would do the absolute best we could by them. I feel very strongly that us behaving very well towards each other in front of them is critical and I never say anything negative about him and he should never say anything negative about me. When we split up, we agreed certain non-negotiable things. So, if the children want us to be together for birthdays and Thanksgiving, and they do, we do that regardless of anything else. That will happen until it is not what they want.

'We said right from the start that this was the habit we were going to form: we will always be a united front. Their dad can come and have lunch with us, and I have been round to his and had lunch with them, and they know that we will behave really well towards each other and do our best for them. If you set the rules right from the start and say you are always going to do it this way, then it is easy and things don't become an issue.

'I don't feel the divorce has caused any long-term damage in terms of my attachment style, my future relationships or how I relate to my children. But I think it would be naïve to say that it hasn't impacted my kids, even though I think I am secure and do my best to provide them with unconditional love. I hope that seeing how well we behave towards each other will help with their attachments to other people, their feelings about entering relationships and how good they can be, as divorce can have a hugely negative impact on children in the long term. Our absolute priority is to minimise that.'

If you decide that breaking up is the best thing for you and your children, there are things that can be done to minimise any negative impact. The first, and perhaps also the most important, is to try, despite all the emotional turmoil and logistical challenges, to be a good parent. By that, divorce researchers mean monitoring your children's behaviour effectively, maintaining authoritative (not lax or punitive) discipline, and providing emotional support and warmth.

This reminded me of our old friend, attachment. If attachment theory is right, making a child feel as safe and secure as possible before, during and after divorce should help buffer the impact of losing, or at least, being distanced from, your partner (who's probably also a significant attachment figure for them).

While this will probably be easier for parents with a secure attachment style, they may be less likely to be divorcing in the first place. But even those who are secure might struggle, as the evidence suggests that parenting can take a nosedive in the first year after separation.

Some parents become irritable, punishing and unpredictable, at the same time as being less loving and spending less time with their children. Others neglect them because they throw themselves into a new relationship* or are flooded with sorrow at the break-up. Others, overwhelmed by guilt, let go of discipline in the hope that this will appease their upset and angry children. And children can be upset and angry. In a study spanning three decades and involving 1,400 families, parents likened their children's behaviour immediately after divorce to Chinese water torture or getting bitten to death by ducks.

Children also tend to do better if, like Cari and her ex-husband, their parents manage to have a positive relationship after splitting up and don't involve them in arguments over living arrangements, visitation and child support. Researchers found that 10 to 18-year-olds whose parents had separated four and half years earlier were more depressed and anxious if one or both of their parents denigrated the other in front of them, didn't allow them to mention the other parent in their presence, or asked them to carry hostile messages to them.

Leaving with respect

Even when we do our absolute best in relationships, including looking inward and taking full responsibility for our actions, sometimes they do not work out. It's easy to think of this as a failure and be overwhelmed by guilt, shame or blame, but it's not as binary as that – relationships, like the individuals involved, are hugely complex and sometimes the right decision is to walk away.

If that's the case, then finding a way to manage anger, disappointment or hurt can go a long way to minimising the damage of divorce for everyone involved – and that

* I explore this in the next chapter.

includes resisting the temptation to draw children into the conflict that so often accompanies the end of a relationship.

This reminded me of something that Chí from Hanoi in Vietnam told me. He was in his early 20s and although he didn't have any children, I suspect his advice is relevant to most, if not all, break-ups. 'I said a lot of mean things when we broke up, I had bad ideas about them and I tried to blame them for everything. You can never forget about that.

'The words that I said hurt them a lot. Now every time they remember me, they remember how bad I was when we broke up. They never remember how happy we were when we came together.

'When everything must end, when there is nothing left, that is when people can really get their emotion out. When I feel like that, when I am full of emotions, I always ask for respect. And I try my best to be respectful.'

Things will get better

When it removes intolerable or interminable pain, the end of a relationship can bring profound relief, but when it means the loss of opportunities, hopes or precious connections with others, a break-up can cause grief and devastation. The emotional repercussions can depend on countless factors, so it's no surprise that the people I spoke to described a wide range of reactions, from emptiness to empowerment, devastation to denial.

'It almost feels bad to say it, but I still feel liberated,' said Cari. 'It was really hard being in that relationship and I only realised just how hard it was when I wasn't in it any more. I like the absence of the things that weren't going well in the marriage. I like that I can make my own decisions – I can listen to Baroque Spanish guitar music followed by boy bands and there is no one to laugh at me. It's not about being single – it's more about not being with the wrong person. If you are a team together, if you are good friends, allies even, and share a lot of values, then you can enjoy listening to music your partner doesn't like without feeling judged for it. We weren't a team for a long time. Life is much easier now than it was before, even as a single mum.'

While some felt relief, others felt resentment. 'Sometimes men want variety and they make a fool of their wives. After our relationship ended, my ex-husband married and divorced three more times,' said Farzana from Pakistan. 'When I was younger, some men would say to me, "You look very beautiful," and I would say, "I have a mirror that tells me what I look like." I did not marry again. I closed the door to men and kept it closed. After the divorce, I realised that the media in Pakistan blames women when relationships go wrong. I decided to start working with Pakistani women. I educated them, developed their confidence and gave them references to get jobs. Most importantly, I told them not to fall in love.'

James from Cornwall, England shared a similar sentiment. 'I was in love. It ended because she cheated on me. I found out through rumours around school, so I dumped her last month. She didn't say much – she pretty much denied it all. I sort of feel now like all girls are terrible.'

For some, their break-up led to loneliness and despair. 'I had not been living in Greece long enough to have many friends of my own, so when we broke up, I had no one to turn to,' said Margaret from Canada. 'If I was lonely before, I was extra lonely afterwards. His whole family went against me and it was as if I had been discarded. I had borrowed a bike from one of their shops and they wanted it back straight away. I flew home, devastated. Some of my friends say that I had a mini-breakdown afterwards, but it was so long ago, it is hard for me to remember.'

Others were almost celebratory at having left. 'You have to be brave, make the decision,' said Pete from the US.[*] 'No one who is happily married decides to get divorced. No one ever thinks, "I'm sorry I got divorced,"[†] and no one ever says it to you either – they say, "Congratulations, you're divorced!"'

Some felt foolish. 'He vanished and then two months after that, I received a postcard from him from Brazil where his now husband is from,' said Miriam from Germany. 'I was so happy that he wrote the postcard! I told my best friend and she said, "Are you out of your mind? The guy left without a word and now after two months he writes you a postcard from Brazil?! There is no reason for you to be happy!" The feeling that I can't trust my own senses stays with me. I feel shocked. I feel completely stupid for not guessing that he was gay.'

Some tried to process their emotions. 'I had to fight for my divorce because people thought that I should stay married, even though he was an alcoholic. It was tough because I was left with two children and no support,' said Aileen from Scotland. 'I have come across a lot of tragedies in my life but when I come up against something, I can't just keep it in the background, forget it and carry on. That seems to be what a lot of people do, but I am the opposite. If I don't face up to things and deal with it then it affects me badly. I am not saying it's been easy, I have had my moments, but you have to have your moments actually. You need to feel and I felt it all, but I have dealt with it.'

Others tried to process their emotions, but failed. 'I'd always been very male about my break-ups – in spite of having people I could talk to, I would try to numb my

---

[*] You met Pete at the start of the chapter.

[†] I'm not sure I agree – the study by Amato and Hohmann-Marriott that I looked at earlier found that around half of people who divorced (those who had happy marriages) were even less happy afterwards.

emotions instead of feeling them,' said Ethan from Wales.* 'So when I broke up with my last partner, I was adamant that I wouldn't avoid my emotions as I had done in the past, because that was why we broke up. I didn't intend to do anything self-destructive – I had a glass of wine, cried for a few hours and one of my friends at the time wanted me to go to the pub with him to try and cheer me up. He didn't know how to guide me through the experience and had his own problems with drinking – I ended up drinking wine measures of gin, without mixers. I barely remember getting home. I had moved back to my family home, so my mother found me lying half in, half out of the downstairs bathroom breathing so little she thought I was dead.'

Some felt the need to run away. 'It was a very hazy time and by the end, I just had to get rid of her,' said Alexandru from Romania.† 'I felt claustrophobic and angry and resentful, and I just had to walk away from that relationship and we didn't talk. I lacked the words to really talk to her about it and it was just "Goodbye."'

Some walked away, more deliberately cutting ties. 'If you realise you're in the wrong relationship, as I did, you have to make a clean break,' said Clair from the US.‡ 'There are too many people who try to be friends and try not to hurt the other person, but the bottom line is that without a clean break, you cannot end something in your heart, you cannot move on, all you do is draw it out.'

For others, cutting ties wasn't possible. 'What really upset me was everyone knew that he was cheating on me,' said Becky from New Zealand. 'After Steve and I broke up, I felt very alone and was a little depressed. Eventually, I

* You met Ethan in Chapter 1. Incidentally, Ethan isn't unusual – research suggests that men are less likely than women to seek support following a break-up.

† You met Alexandru in Chapter 3.

‡ You met Clair at the start of the chapter.

changed my life, went back to studying, got a job, met my husband. Life was brilliant and I was happy. Until my sister rang me and said, "I'm in a relationship with Steve."'

A sense of loss

Despite the variety in my interviewees' experiences, there's one thing they all shared, even those who had comparatively happy break-ups – a sense of loss. Not just the loss of a partner, but also friendships, status, safety, security, meaning, home, neighbourhood, children, an imagined future, intimacy, income, ideology or identity (among other losses).

'It wasn't just the breakdown of a relationship – it was a collapse of the way I understood the world,' said Felicity from the US. 'It wasn't just a separation from someone I believed to be "the one" – it was a whole belief system that I had grown up believing in and really treasured as "the answer". I believed that you would meet someone, they would see you, you would see them, you would both be enough for each other, and you would decide that was that. Then it crumbled around me and my ideology didn't make sense any more. The grief of the loss of the ideology was as powerful and as present as the grief of the loss of the relationship.'

'I didn't just break up with my husband,' said Sarah-Jane from Canada. 'I broke up with his friends and family. When the marriage dissolved, our mutual friends were no longer interested in being my friend because they felt compelled to take sides. I still see them around and they act like we have never ever known each other. It's awkward. I will be friends with absolutely anybody unless they don't want to make an effort and be part of my life, or unless they are very negative and don't treat me very well. When that happens – as it did with our mutual friends – well, I hope the door doesn't hit their ass on the way out.'

'Break-ups are a paradox,' said Ethan. 'What do you do when you can't talk to the one person in the world that you are supposed to talk to about a devastating loss? This is the person you chatted to about all your intimate, private issues and they're gone – previously you could have asked them what to do or enjoyed the catharsis of being listened to, but now you have to resolve these problems by yourself. As a male, I found that particularly hard. And as I looked around at my friends who had moved around the country, I found myself thinking, "Who's left?"'

As humans, we haven't evolved to operate independently. Instead, we're programmed to develop attachments in order to survive – as an infant, it's literally a matter of life or death. This means that the attachment system is incredibly powerful, capable of driving our thinking, memory and behaviours – so it's no surprise that when we separate from one of our key attachment bonds, we suffer.* But as Ethan pointed out, not only do we suffer the loss of someone who may have been vitally important in our lives (as well as a host of other losses), we lose the very person we would have turned to in such a crisis.

But romantic partners aren't the only attachment figures we have in our lives – close friends, siblings, parents and other relatives can also provide the support and love of an attachment bond,† and these relationships can be key to coping with a break-up. In a review of 21 studies looking at divorce, researchers found that people spending time one-to-one with close friends or family

---

* Assuming that they were an attachment figure, of course. Break-ups can also temporarily or permanently change our attachment system, which I will explore in the next chapter.

† Although research suggests that family support isn't always a good thing: in the Swedish Level of Living Survey, a long-term study following 0.1 per cent of the Swedish population aged 18–75 over a 10-year period, researchers found that divorced people who often saw friends had higher well-being. But those who often saw relatives did not – in fact had lower psychological well-being.

could protect people not only from symptoms of depression, anxiety and stress, but also physical symptoms following divorce.[*] But it's not only close relationships that can be helpful – people who spent time with circles of friends (or support groups) were happier, more positive and better at coping, and had higher well-being and life satisfaction.

Animal studies reinforce the finding that friends and familiar faces are essential sources of support (well, in the context of stressful situations – as far as I know, sheep and rats don't formally 'divorce'). Sheep are less distressed when they're put in a new pen when accompanied by someone from their flock, rats are less afraid of electric shocks if they're with a pal and squirrel monkeys are less stressed when separated from their group if familiar adults are with them.[†]

Having close, empathetic and positive relationships to fall back on is the ideal. But for some, it's only when relationships end that they realise their social support networks aren't quite what they'd hoped they would be.[‡] In this case – and

---

[*] Although I suspect the specific relationship is an important factor – spending one-on-one time with certain people might have the opposite effect. Incidentally, neuroscientific research suggests that we don't even have to be in the presence of people we love; we just have to remind ourselves of them: when people were shown photos of a loved one or their own children, this was enough to deactivate the amygdala, the brain's smoke detector. (So, if you have photos of your ex anywhere, it might be an idea to replace them with a photo of someone else – preferably someone you love.)

[†] Researchers found that it took a number of familiar adults, or just their mother, to calm them down – suggesting that the quality of the relationship is important.

[‡] Although it's not just about having the right people in your life – getting support often requires an ability to ask for and receive it.

even for those who do have sufficient back-up – research suggests that we can also try to offer ourselves support and love in the form of self-compassion.

Self-compassion

People who are self-compassionate have healthier physiological responses to stress, they're happier, more optimistic, motivated and satisfied with life, and they have lower levels of depression, anxiety, stress, fear of failure and shame about their bodies. They also find break-ups less upsetting: in one long–term study, researchers found that people who were more self-compassionate were less distressed by their divorce up to nine months later, even when a large number of other factors were considered, including their attachment style.

So, what is self-compassion? Psychologist Kristin Neff argues that it consists of three components. The first is self-kindness, which involves being understanding, tolerant and encouraging towards yourself rather than being disparaging or cruel. It also means providing comfort and warmth when things go wrong, rather than ignoring your pain or simply trying to fix the situation.

The second component is the recognition that everyone makes mistakes, behaves badly and fails, and that we are all, as humans, imperfect and subject to suffering. Instead of assuming that everyone else is having an easier time, people who are more self-compassionate recognise that challenges, failures and loss are simply part of being human. This was something Sarah from the US tried to keep in mind: 'I read about other people's experiences and it put things into perspective. It gave me a sense that, no matter how devastating it is, this feeling will pass, things will get better at some point and there are a million people out there that have experienced the same thing.'

The third component is mindfulness. The aim of mindfulness is to develop the ability to observe your experience in each moment – and to do so without judgement, avoidance or repression.* This is relevant to self-compassion because by paying attention to thoughts, feelings and physical sensations, you will be in a better position to notice when you're suffering (rather than automatically bury it, for example) and less likely to make it worse (with, say, self-criticism or self-attack). So, you might notice that you are someone who, in this particular moment, feels disappointed that their relationship didn't last – rather than get caught up in your thinking and see yourself as someone whose life is disappointing or who is a disappointment.

This idea of self-compassion is something that came up in a number of my interviews. I met Jenny on a flight to Zurich, Switzerland. She spent most of the journey staring out of the window, clutching a blunt pencil and a tangerine notepad. 'When I split up with my boyfriend, it felt as if my limbs were being slowly sawed off.' My illusion was shattered: I had seen this guy as a complete guru, but he turned out to be a bully. I put a stake in the ground and I walked away from him, but it shattered my world. The next day, I couldn't get out of bed. It was like the most horrific hangover, but I had only had one glass of wine. I was wiped out. I couldn't move. I was having a breakdown. I went kaput.

'I had to redress the balance of how I related to others and to myself because I was going to end up in hospital, basically, having emotionally bankrupted myself. So I did a life audit. I reviewed everything on paper: all my fears,

* I don't mean trying to not be judgemental in the first place, only to notice, without judging, what you think or feel about yourself or others.

resentments and the story of my life. Through that, I found the patterns that I was living.

'I've slowly woken up to the fact that I need to be selfish enough to take care of myself. If I'm well slept, restored, spiritually fed, then of course I'm going to show up emotionally for other people. If I'm depleted inside, if I have abused or abandoned myself, I have nothing to give.

'Culturally, we're not taught about loving ourselves. We're taught self-sacrifice. But if you're self-sacrificing to the detriment of your own emotional, spiritual and physical health, that doesn't help anyone. We're on a plane right now, and it's the same as putting the oxygen mask on yourself before you put it on others.

'I find mindfulness and compassion important. It's easy to think, "Who cares about meditation in the morning, I'll just lie in bed all day and then go and eat shit for lunch." There is a time and place for that, obviously, but in the long term, it's important to build that 'love' muscle in your mind.*

'Human beings are imperfect and will fail my expectations at some point. And that means I'm allowed to mess up sometimes because I'm not perfect. Attacking myself doesn't work. I can try and attack myself to make me perform better in life, but it doesn't actually make me perform better. What actually works is gentleness and compassion.

'Accountability, yes, but always with compassion.'

---

* Jenny was right – there's a growing body of evidence that the brain does in fact operate as a muscle and can be strengthened with exercise. For example, a number of studies have shown that eight weeks of mindfulness training is enough to cause structural changes in the brain.

CHAPTER FOURTEEN

Borrowed People

'We were married 65 years and 49 days,' said Maurice. Having long admired his poetry, I was excited to travel to his retirement flat by the sea to speak to him. We spoke surrounded by his daughter's artwork, overlooked by a black-and-white photograph of his wife Olive on the mantelpiece.

'I think that, in the last few days, Olive knew that she was dying. But I didn't believe that she would be. She went into hospital with a hiatus hernia and she ingested some of the fluid that was coming back up into her lungs. That is how she died. To me, she was going to come out in a few days when she went in.

'When I walked out of the hospital, I found it difficult and I stopped by one of the bins and rested my head. I didn't want to walk any further or go any further, and I didn't want to live any longer.

'We scattered her ashes in a meadow garden on the cliffs where we lived. I used to make myself walk there and walk around the garden, talking to her, weeping aloud and telling her what was happening with the family, and then go back home. I had hoped that one day I would flop down there and that would be it.

'It occurred gradually to me that the friendships that we had with other people were the important things and the things that were left. And it was friendships and family love, and the togetherness that we had, that helped me get over it. It was never an easy time at all, it was lonely. Going back into the house was lonely, but going out was also being alone. It was a very difficult time.

'It was unbelievable and unacceptable and surreal. That is all I can remember. Looking at other couples arm in arm or hand in hand was extremely difficult. It was sort of not happening to me, I was a different person. I couldn't face it. I look at her photograph and I still talk to her now and again. I think about her a lot and I think with her quite a lot.

'Nobody can cope with these things, but they do. Human nature is remarkably resilient. I mean, what am I doing here at 95 years of age, still thinking that I will go buy some potatoes this morning?

'I think what made our marriage last was the feeling of support and friendship that we got from each other against some of the difficult things that happen in life. Support and reliability, boring as it might seem, and regularity, dependability. That was it. We enjoyed being together. There is always the feeling in the end that one could have tried harder. I grieved for the loss of the companionship that I had and I fell on to the shoulders of family.

'I am engaged in my family life, my children are mourning the loss of their mother, and we share it. I need to grieve but I don't make a big thing of it. When it takes me over, it takes me over. I don't often sit and cry now, but I used to if something triggered it. You have to let it go and let the other person's absence take you over. Immerse yourself in it totally because then you can begin to come out.

'It is never going to change. Olive is never going to come back. But in my mind, soul and body, she lives on. Olive would always iron the bedsheets and the pillowcases and when she made the bed, the wide hem would be up at the top end. The sheet would be folded in a certain way and laid out on the bed. So, I still do that. That is a ritual isn't it? Adding milk to tea and not tea to milk. Those little things. And we very often sing together, Olive and I.

'Quite frequently I make a meal, sit here in the same place and eat it – and at the end of it, if I have enjoyed it I say, "Well that was good, wasn't it Olive?" and I answer myself, "Yes, Maurice, it was."'

I struggled to hold back my tears during my conversation with Maurice. Not only was I devastated by his loss, I was full of sorrow because my grandmother, who was like a parent to me, had died a few weeks earlier. Maurice's words brought a tiny chink of light into the dark emptiness that took up so much space inside me, I wondered whether there was room for anything else.

Grief is difficult to describe because it's as individual as the relationship devastated by death. (Although it can – and usually does – follow both death and divorce, I'm going to focus on grief in the context of bereavement.*) Losing a partner can mean very different things to different people. It might mean losing a best friend or a bed warmer, a cleaner or a carer, a co-parent or a cuddler. Or it might mean losing your accountant, assistant, advisor, abuser, decision-maker, dishwasher or lover – or all of the above.

There's still a lot we don't understand about grief: do people who have the best relationships suffer the most when their partner dies? Do high-conflict relationships lead to more sadness when they end? Are people who don't seem to grieve repressing their emotions? Or are they resilient? Or not that attached in the first place?

What we do understand – or at least, what most psychologists seem to agree on – is that we don't experience grief in stages, like denial, bargaining, apathy, anger and acceptance.† Instead, there's a huge amount of variety: for some, grief can tear down their sense of meaning and

---

* The word 'bereavement' refers to experiencing the death of someone close to you. Defining grief isn't as straightforward as people have different definitions for it – but on the whole, it's used to describe any form of distress that is a response to loss. I include phrases like 'lost' or 'passed away' just to have variation in language, but I don't think the language around death is ideal. As Jennie said: 'You haven't lost them at all. You know exactly where they are.'

† This is Elisabeth Kübler-Ross' five-stage model. Although this view was relatively popular in the past, there's very little evidence for it.

belonging like an emotional avalanche, for others, it's barely noticeable.

Acute grief

Grief is usually (but not always) more intense to begin with and in the first few months can entail a veritable maelstrom of thoughts, feelings and memories, including disbelief, longing, sadness, irritability, anxiety, guilt, fear, anger, shame – and more. When researchers interviewed gay men four weeks after the loss of their partner, they expressed no less than 22 different emotions (14 negative, 8 positive).

In a US study of 350 people who had lost their partner two months earlier, 87 per cent felt a piece of them was missing, 70 per cent found it hard to believe that their partner had actually died and 20 per cent found it difficult to concentrate. On top of that, 59 per cent felt lonely – and for most, like Maurice, this sense of loneliness continued even when they were around other people. In another study, 72 per cent had involuntary memories of their partner, 40 per cent felt numb and 31 per cent had trouble sleeping.

'It was a weird hollow feeling,' said Chloe, who lost her husband after a decade of marriage. 'It was a blackness, a heaviness, not being able to see anything positive. It was very low and I was functioning and doing the things I had to do to keep going, but I was doing them on autopilot. I was just going through the motions and I found no joy in anything. It was a dream-like state where nothing felt particularly real. I forgot about all the good times. All I could think of was the last few days of his life, that is all that kept replaying. My memories are quite foggy, but I would describe it as a deep despair.'

'Early grief is … tough,' explained Andy, who was in his early 30s and whose partner had died five years earlier. 'Lost. That's another word I would use. There will never be a word that is big enough to describe how I felt. I remember picking up my phone and hearing, "Have you spoken to Aaron? He's had a car crash and he's not here any more." He was 27. I remember hanging up and dropping to the floor.

'In those first few months, the only thing I could manage to do was go to work and go home. I was skipping meals, missing messages, missing calls. My mum was really worried as my face became very drawn; I wouldn't eat for days. I stopped doing everyday things that should be done. For days at a time, I didn't brush my teeth.

'I felt nothing – I felt no emotion for the people who were showing me love, care and attention. They couldn't bring him back, so I pushed them away. I just didn't want anybody around me, I wanted to be on my own. Looking back, that was probably the worst thing to do.

'It took me about eight or nine months to snap out of it. The worst thing about it is that I didn't even know eight or nine months had passed. I look back and I think, "Wow – that was almost a year." I don't know what changed. At some point, I realised that I needed to start living again. Because I wasn't – I was surviving.'

Acute grief can look very similar to depression and can trigger it. A long-term study found that people over 65 suffered 75 per cent more depressive symptoms in the first few months after bereavement compared with before. And a review of 11 studies looking at 3,481 people who had lost their partner found that 22 per cent were diagnosed with major depressive disorder during the first year after their loss.

Depression and grief have a lot in common: they can both feature intense sadness, appetite and sleep disturbance. But there's one important difference: grief doesn't usually lead to the same loss of self-esteem as depression. Psychoanalyst Sigmund Freud believed that, in grief, the world looks poor and empty, whereas in depression, the person feels poor and empty.

As well as depression, the death of a loved one can also cause PTSD. Symptoms of PTSD include flashbacks, nightmares and intrusive thoughts, irritability, detachment and avoidance of certain activities and situations.

If the death was unexpected or traumatic, as in Andy's case, PTSD can be relatively common. In a US study of 350 widows and widowers, 36 per cent of people whose partners died from suicide or accident met the criteria for PTSD two months after their loss.* It can also occur in other types of bereavement, albeit less frequently (around 12 per cent according to the review of studies I mentioned earlier).

'Integrated' grief

For most people, acute grief eventually turns into a calmer version of itself, and people find a way to move their attention from their partner to both their future and the world around them. Psychologists don't agree on whether this transition requires some form of processing or 'grief' work, or whether this shift unfolds naturally.

This calmer version of grief isn't static – it morphs and melds over time, with sadness or longing bubbling up on anniversaries, family holidays, major life events or losses. What I can say is that it can last a long time. In fact, it may never end.†

'It has been nearly three years and I still miss Graeme,' said Suzy from Sydney, Australia. 'It is easier now. I can

---

* That doesn't mean that 36 per cent of all widows or widowers *everywhere* will develop PTSD: as well as involving a high number of elderly, middle-class females, this study only included people in San Diego county who were willing to allow researchers into their homes at a time of great turmoil.

† More specifically, it's normal for this calmer grief to linger (albeit getting easier with time). But for some, it's the more dominating, overwhelming and acute grief that persists, and time does nothing to help – this is what psychologists call 'complicated grief', which I will come on to later. People suffering from complicated grief may need the support of a therapist to overcome it.

remember at first it felt like I was going to drown, like a wave was going to hit me and I would drown. That is the only way I can describe it. Now I very rarely feel like that and certainly not as intensely. I will have moments of panic, but it won't be that overwhelming or feel like a tidal wave. I have got a lot better when I think of it like that.

'My regrets have definitely decreased. I couldn't go back in time, so I beat myself up about everything, especially about what I didn't do.* Thankfully that did gradually fade because rational thought took over to a degree. And I just couldn't live like that, with those regrets at that level. Somehow I started beating myself up less and tried to concentrate on what I did do rather than what I didn't.

'I still think about him every day, but probably not quite as much as I did. In the beginning it was awful because I would wake up expecting to see him and he wasn't there. That was hard. I still wake up a lot of the time thinking about him.

'I talk to Graeme still. Sometimes, in my head, I can hear him saying something really dry and funny and I will laugh because I know that is exactly what he would have said. That is brilliant because that feels like he is never going to go completely. I hope he doesn't. I don't think I will ever not miss him.'

The never-endingness of grief

I often wonder whether many people (at least in Western societies) have completely the wrong idea of grief – namely, that it's a relatively short, and finite, process, a bit like falling over and grazing your knee. It will scab over if you don't touch it, it will heal and you might have a small mark, but that will disappear too and then you'll be back to normal.

---

* Psychologist Richard Gardiner makes a related point – he suggests that the regrets we might have after someone we love dies are an attempt to control the uncontrollable.

But losing your partner, one of your main attachment figures, is more like discovering that gravity doesn't exist and then having to find a new way to live life without that grounding.* When one of the most important people in your life dies, you don't just lose someone 'out there', you face your own mortality and aloneness, as well as losing a part of yourself and a set of basic assumptions about the world, particularly when that someone was one of the few people who gave you a sense of safety and security. So why would the grief borne out of such profound loss be short-lived?

In many societies around the world, death rituals are elaborate and extend over weeks, months or years – far longer than is common in Western societies. They might require the people mourning to wear special clothing or markings, to tear their clothes or skin, abstain from washing, beat themselves or shave their head.

The Kota people from the Nilgiri Hills of South India, for example, have two funeral ceremonies: the 'green funeral', which happens shortly after the body is cremated, and the 'dry funeral', which is held at least every year and includes all deaths since the celebration of the last one. The difference between life and death isn't black and white to them, and they believe that dead people retain a number of earthly qualities that only end when the second funeral is carried out. A widow is married to her husband until the end of the dry funeral and if she gets pregnant before then her pregnancy is attributed to him.

Meanwhile, the Toraja of Indonesia mummify their deceased with a solution of formaldehyde and water, and a mummified loved one isn't seen as dead, but is considered to be a *to makula*, or sick person. They often live in the house with their family for long periods (months or years), during which time they're fed, dressed and talked to. When they are eventually buried, they're brought back out every

---

* I owe this comparison of bereavement to sociologist Peter Marris.

few years to be cleaned, given fresh clothes and have their crypts spruced up.

As I researched the Kota and the Toraja, I wondered whether the slower pace of their death practices and beliefs were closer to my interviewees' experiences of grief than the buttoned-up rituals of many Western countries.* The people I spoke to were explicit: grieving the loss of a partner was not a short process. In fact, many doubted it would ever end. And the evidence backs them up.

In the Americans' Changing Lives study, researchers interviewed thousands of people who were representative of the US population, 768 of whom had lost a spouse anywhere from three months to 60 years before.† When deciding on how to carry out the study, researchers debated who to ask about their loss and some were reluctant to ask questions 10 years or more after bereavement. In the end, they did, and in doing so, they were surprised to discover that grief lasted for decades.

Among other things, the study looked at anniversaries related to their bereavement – for example, anniversaries of their loss, their partner's birthday or other important dates. Researchers asked people whether, on these anniversaries, they ever felt the sadness and loneliness they felt immediately after their partner's death. On average, it took about 53 years

---

* That's not to say that cultural practices involving long-term mourning are always ideal. Traditionally, South African Zulus practice *ukuzila* for a year after losing their spouse. They wear black or blue clothing, shave their head and behave in a way that means people know they're grieving (for example, giving way to others, avoiding eye contact and bowing or bending their knees). Men aren't expected to practice *ukuzila*; women, on the other hand, are – even if their husband left them for many years to be with other women or was abusive.

† To be specific, it was representative of non-institutionalised people aged 25 or older and living in the continental US.

for these 'anniversary reactions' to almost disappear. Fifty-three years.*

A debilitating cost

Many have claimed that grief is the cost of love. If it is, the evidence suggests that it can be an exceptionally high cost, one of the worst imaginable. In fact, it can be so debilitating that it makes a life of isolation seem relatively appealing by comparison.

But something Maurice said made me reframe my understanding of grief completely. He told me that he no longer saw his grief, or its enduring nature, as negative – instead, he saw it as a confirmation that he had loved, and may never stop loving, his wife very deeply.†

'The grief has lessened ever so much, and I accept it, I am not afraid of it,' he said. 'I need to grieve but I don't make a big thing of it. When it takes me over, it takes me over. It does so less frequently now, and less deeply. It will come out at times and I grab it because otherwise nothing would have been worthwhile or true. It was real.

'It becomes acceptable with gratitude because it confirms to me that what I, and what I believe Olive, had valued of our marriage is right and true, that I am not making things up, including the difficult bits.'

Ending things before they begin

'In the third year of being together, John became ill unexpectedly. We married on May 2nd and he died on

---

* Anniversaries aren't the only time that people felt their loss. Twenty years after their bereavement, people spoke about their partner once a month and still thought about them once every week or two on average.

† Although in the case of complicated grief, enduring sorrow might also be the result of unhelpful thinking.

May 26th,' I met Jennie on a children's book illustration course. She had just joined a local choir and as she spoke-sang to me, her melodic voice disguised the pain of what she was describing.

'To outsiders it was like, "Oh, what a tragedy," and it was. It was not easy, I didn't enjoy a lot of that process but would never have changed it. It was just such a glorious relationship to have in my life. I wonder how different life would be now if he had lived.

'Immediately after he died, I was in complete disbelief. I would see him everywhere; somebody's hair or clothing would remind me of him. There was one guy I almost stalked over a couple of months because he was so similar to John physically that it was like he could bring him back to me. That was quite painful, quite weird. Fortunately, that wore off and I came through that.

'For three or four years, I went a bit wild, trying to see if I was actually alive because it was not easy. I had a couple of completely unsuitable flings. I had a one-night stand with a guy who was so physically attractive, a brilliant-looking guy, irresistible. It was a sort of madness, which I knew wasn't going to work. Turns out he was married, so I ended it.

'I had a couple of completely unsuitable relationships, but I knew that they were not the kind of person that I could fit into my life because, by that point, the children were small, and we were three people, we came together, we were a package. Okay, I could have a night out with somebody, great, but ultimately I was part of this group, so if I was ever going to form a lasting relationship, it would have to be with somebody who would love us all.

'I didn't want anybody, but it just happened. I remarried. Totally different kind of person. Except he and John shared one quality, a quality I think I also have: they were quite secure. It's the quality that I think made the relationships work.

'We just soldiered on, we fought and he fought, but he didn't make it, he had liver disease and he died. I have no idea where you find the reserves from. Well I do, I think you borrow these people don't you? And what they give you in your life – it carries you through.

'Having lost somebody twice, I know that something wonderful can end – and that can sometimes put me under pressure. I find myself thinking that I have to know whether a relationship is going to work, make the most of it, but squash it all in because it might end. There's a sense of panic.

'What I'm trying not to do is force it, or let the fact that the other beautiful experiences ended ruin this one I'm in now. I try to be calmer and more able to live with myself appreciating that it's wonderful being with somebody else in so many ways, sharing so many things. But I do recognise that if I'm not careful, having experienced those losses, I could make something end before it's had a chance to begin.'

In the distance, we heard the faint wail of an ambulance siren, its cries growing louder and louder until it shrieked past us. It felt somehow appropriate that we'd been interrupted by something representing the fragility of human life. As the siren began to fade away, I reflected on Jennie's sense of panic and her observation that, having been bereaved twice, she could make something end before it had even begun. And this brought me back, once again, to attachment theory.

In an ideal world, attachment relationships provide us with a secure base from which to explore and a safe haven in times of crisis. When something stressful happens, the attachment systems of secure or anxious people are activated and drive them to get close to their loved one, who would ideally calm them down. (Avoidant attachment systems are less straightforward, but I will come on to that.)

The problem with bereavement (and break-ups) is that, as well as suffering the loss of someone you love, you're deprived of the very person you would turn to in the face of such a crisis. But that's not all: not only do you lose your safe haven in the face of disaster – you also lose your secure base from which to explore. And this is why some researchers argue that bereaved people can lose interest in the outside world.*

Anxious people tend to spend more time thinking about their partner and are usually particularly keen to be close to them – but when their partner dies, these tendencies can make grief much more painful and longer-lasting. There's very, very little research on attachment responses to bereavement (and even less that relates specifically to the loss of a partner), but what there is confirms this.

In a study conducted two weeks after the Gulf War, researchers interviewed people who lived in the area targeted by most of the Iraqi Scud missiles. People with anxious attachment styles had more anxiety, depressive symptoms and psychosomatic complaints than those who were secure.† And in a long-term study of 438 parents who lost a child, researchers found that those with an anxious attachment style suffered more grief and symptoms of depression than those who were secure.

In some cases, the preoccupation that anxious people can have with their loved one can result in what psychologists call 'complicated grief'. This is essentially where acute grief doesn't calm down, but instead persists in all its longing, devastation and sorrow. Typical symptoms include: insistent, intrusive thoughts about their partner, inability to accept the death, excessive guilt, intense anger, avoiding people, places or reminders of their loved one, extreme longing, difficulty finding meaning in life, and

* Although this may also be down to the barrage of platitudes and avoidance they would face if they did venture out …

† Psychosomatic complaints are when psychological distress manifests in physical bodily symptoms. (Although I should also point out that some 'psychosomatic' symptoms can also have non-psychological causes.)

feeling that they could or should have prevented the loss. In one study, people who had an anxious attachment were more distressed 14, 25 and 60 months after the death of their partner.*

Instead of trying to get close to someone when something stressful happens, avoidant attachment systems shut down, deny the need for others and emphasise independence. This means that when a highly avoidant person loses someone they love, they're unlikely to try to think about their loved one – instead, they're more likely to repress emotions, be self-sufficient and deny any need to be comforted.† Perhaps that's why some studies have found that avoidant attachment isn't linked to high levels of grief.

But it's not as simple as that. Avoidant defences might work for minor stressful events, but when the trauma is significant or ongoing (as is often the case when you lose your partner), there's some evidence to suggest that avoidant defences can collapse, leaving those who are avoidant to behave in a very similar way to anxious people – or worse. In the long-term study of 438 bereaved parents I mentioned earlier, those with an avoidant attachment style suffered

---

* Unhelpfully, this study only measured attachment for the first time 10 months after loss, by which point death may have caused a shift in attachment style. And that makes it harder to know whether anxious attachment played a role in causing complicated grief.

† Some argue that, if someone with a highly avoidant attachment style had managed to shun intimacy and be fiercely independent when their partner was alive, they might have less of a relationship to mourn. But I suspect this is a pessimistic view. Highly avoidant people may show less outward signs of grief as they are not great at labeling and being aware of their feelings – instead they have learnt to dissociate from them and are less likely to want to express grief. But that doesn't mean they don't suffer. They might, for example, have more psychosomatic complaints.

more grief and depressive symptoms than those who were secure.*

There are some researchers who argue that avoidant people are resilient – but I'm not convinced. They might appear resilient, but that may be because they're disconnecting with their needs and burying their sadness – leaving their bodies to tell a different story.

Avoidant people don't usually talk negatively about their parents – in fact, they're often very positive. But when researchers carried out 'skin conductance' tests, they found that, while avoidant people spoke positively about their parents, their bodies had a very different emotional reaction.† Meanwhile, in a study of bereaved adults across Arizona and Colorado in the US, avoidant people didn't have high levels of grief or depression, but they *did* have more psychosomatic complaints.‡ Although avoidant defences might block *conscious* access to anxiety and depression, they may not be able to block more subtle and subconscious reactions.§

---

* Research beyond bereavement has found similar results. In a long-term study of mothers whose children were born with congenital heart disease, avoidant people experienced the greatest deterioration in mental health (and marital satisfaction) over the seven years following their child's diagnosis. Mothers' attachment insecurity at the start of the study (whether anxiety or avoidant) was also linked to their children's emotional problems and poor self-image seven years later.

† 'Skin conductance' is when skin momentarily becomes better at conducting electricity as someone is 'physiologically aroused'. The idea is that, if they're physiologically aroused, it's likely to be thanks to emotion.

‡ And in the Gulf War study I mentioned earlier, those with an avoidant attachment style had more psychosomatic complaints than those who were secure.

§ That said, ideally we would have more research on this in the context of bereavement to reach more definitive conclusions. I should also add that it may be difficult to tell the difference between successful but very active suppression and, to the extent an avoidant person avoided intimacy and interdependence with their partner, a lack of anything to suppress.

The other way around

While your attachment style can impact the way you react to bereavement, it also works the other way around: bereavement can impact your attachment style. Although attachment styles tend to be pretty stable over time, major life events, such as the death of your partner, can shift them – or at least that's the theory. This shift might be temporary, so someone might, for example, develop an unresolved attachment style in the short term until they are able to process their loss.* Jennie's description of going wild and having completely unsuitable relationships and flings immediately after John's death made me wonder whether she went through a brief period of this.

But the change can also be long term, shifting from secure to anxious or avoidant (or unresolved) for much longer than a matter of months. I wondered whether this had also happened to Jennie, as the sense of panic that she described, together with an acute awareness that relationships can end, made me think of anxious attachment. It would certainly make sense – if not one but two of your partners passed away, why wouldn't you be more aware, and perhaps even scared, that if you met someone else, the relationship could end at any point?

There's very little research on the impact of losing your partner on attachment, but studies of children suggest that loss is linked to a shift from security to insecurity. A long-term study that followed people from one to 20 years old found that those who experienced the loss or life-threatening illness of a parent, parental divorce or had themselves suffered abuse, maltreatment or a life-threatening illness, were twice as likely to shift from secure

* Attachment researchers use the word 'resolve' instead of process. And confusingly, what they call 'unresolved' in adults is what they call 'disorganised' in children – so I stick to 'unresolved' for both.

to insecure as those who hadn't. Unhelpfully, this, and the handful of studies that do explore shifts in attachment style, unhelpfully group loss with other stressful events – so when there was a change, it's hard to know whether it was specifically down to bereavement or not.

As I reflected on Jennie's interview and everything I'd read, I remembered something that I explored in Chapter 3 – that insecurity stems from a fear of loss. Different attachment styles deal with that fear in different ways – highly anxious people respond by clinging on to relationships for dear life. Highly avoidant people, on the other hand, respond by denying their need for others and closing themselves off – after all, if you don't love in the first place, you can't suffer when the relationship ends.

There's nothing wrong with people who have an insecure attachment style – their brains have simply learnt a lesson and generalised it in an inappropriately broad way (which is how much of our early learning works). Our brain's first job is to keep us alive. Overreacting to possible threats is a valid, and often successful, way of doing that. It may not be helpful for grieving once a partner dies, but that is less important in the evolutionary scheme of things.

It made perfect sense to me that after losing a partner, you might be more sensitive to relationships coming to an end. Or, alternatively, you might try to be compulsively independent in an attempt to avoid having to lose someone again (which was my approach after my father figure died). Both are just learnt reactions to what we've experienced in the past – reactions that unfortunately make it more difficult to have loving, intimate and committed relationships. But while they may be largely subconscious and hugely ingrained, there are still things we can do to change them, if that's what we choose to do.*

---

* I talked more about changing your attachment style and 'earning security' in Chapter 3.

I approached Jim in an international airport in the US. He was staring out of floor-to-ceiling windows at planes preparing to take off and land. He was in his 60s, with short, charcoal hair, and I honestly thought he would say no to being interviewed, but he very quietly obliged.

'It's very difficult when someone you love passes away. The person I loved died. She passed away. She's been gone for 12 years. Other relationships that you have with other people help – like my relationship with my daughter.

'You move on, time heals a lot of things, you develop another life. It is all surreal. It seems like a different lifetime – that was one lifetime, this is another lifetime, and I am sure I will have other lifetimes too, you know, within this lifetime …' He trailed off, staring into the middle distance for a while, and we sat there, amid rows of empty seats, in silence.

'But I think it is possible to love again after you lose someone. I don't think love is limited. I don't think you have a finite amount of love so that if that person dies, you can't love again.

'I think you have an infinite amount of love, whether it is love of a person, love of nature or being alive. You can continue to love, you can re-love.

'I would just say don't close yourself off, be open to love. Know that love is out there all the time and it is there for the grabbing. Don't close yourself off to it.'

Trotting out platitudes

'One of the things that I do struggle with is when people say, "It's all mapped out,"' said Jennie. The ambulance siren had disappeared into the distance and we'd taken a break from love to talk about children's illustration.

'I don't get any comfort from that and I don't want to be on a road where I am heading towards something that is lined up. I got hundreds of these sorts of comments. Well, when people weren't avoiding me. As if people don't realise that any hurt they can give you is nothing compared to the hurt that you are already feeling.

'You know those cultures where they wail? I thought a lot about that. There's something about the public cultural display of grief, that communal thing, the sharing – I wouldn't have minded that. I did my wailing in private. I would have liked the public wailing.

'I wanted people to say how lovely he was, because he was not there and I was not sure if he ever was. I wanted to hear that because it brought him straight back. It might make me cry, but it's wonderful. But people don't like to do it. You just want to hear their name as it brings them into the room it means they existed, it was real. It's very odd and I wish people would have the courage, confidence and empathy to say the name of the person you've lost.*

'"He had a good innings." That's another one. Even the idea of a good innings is bollocks. That's just awful. What difference does that make, whether he had a good innings or not? Makes no difference to me, mate. People say, "Oh it's only two years since he died, it's early days." Please don't trot out that stuff, because what does that mean? Mary, my good mate, whose husband died eight years ago, to her it is still early days. So just don't trot out the platitudes.'

Trot out the platitudes: I loved the phrase. It seemed to perfectly capture the reactions to death (and divorce) that

* I particularly like what Maurice had to say about this. 'I don't back away from other people's deaths and I am not afraid to say, "How are you coping now without Roy" or whoever. They probably weep but they thank you for it as they need to talk about it. In the war, you lost people all the time – you felt sad and you accepted it because it could be you next. This was widespread throughout the world at war. Anything that helps to share that death is important.'

my interviewees had told me about, reactions that seemed intent on denying, or at best minimising, the pain of loss. Reactions that treated the death of a partner and the ensuing grief as unspeakable, unimportant, something that fell to the bereaved to deal with, to lock away so that others wouldn't have to face its inescapable horror.

There's some evidence to suggest that, instead of pulling people closer, the sorrow of grief does the opposite and seems to drive people away, at least in Western societies. When US students read transcripts of people describing the loss of their partner, they were less likely to comfort, and more likely to avoid, people who were more negative about their grief. Admittedly these were strangers reading transcripts of people they'd never met rather than old friends speaking face-to-face, but I think it says a lot about how the 'they' expect people to behave in the context of sorrow and loss.

'Do you have any idea how many times I have heard glib one-liners?' I met Adeleh in a small clothing store in Taipei, Taiwan. 'My favourites include, "It will be okay," "Things will turn out alright," "You have to be strong," and "Everything happens for a reason." I heard these from otherwise very intelligent people that I could count on – or that I thought that I could count on. The one-liners were then promptly followed by a subject change.

'I wanted to shake people. I felt insulted by that response. I felt like saying, "Can you accept the fact that part of why this is stressing me out is that I'm facing my own mortality a lot sooner than other people do? It will all be alright. Why? You don't know whether it will be alright, so why don't we talk about it together because that would really help me."

'There are people who can have nothing by way of words to offer, but I know that they are 100 per cent there and just want to be around in case I need anything, and that makes me feel good. Whereas there are others who make me feel worse than I already do, and I think, "You are using oxygen that I could do with, get out of my house." I just stopped talking to people. I decided that I would figure it out on my own instead of having to listen to the rubbish that other people were spouting.

'There is sadness because you are innately sad, and there is sadness because you feel like you are feeling something that others don't – and they're two different things. It's that sort of societal sadness that could be avoided if we were all much more honest with each other, particularly when we lose people that we love.'

What struck me about Jennie and Adeleh's experiences was the isolation of grief. Facing the death of your partner is isolating as it is. It's a reminder that, no matter how hard we try to live life two by two, no one can die someone's death with them or for them. It's also isolating because you lose your companion in life, the person you talked to about the intricacies of your day, the person you shared food with, the person who comforted you when you felt small. But the people I spoke to felt even more alone as a result of the way others responded, and I wondered whether this was an inevitable part of grief, or whether there was a better way of supporting the bereaved.

Anthropologists who have observed bereavement, grief and mourning around the world argue that, while some cultures (mainly in the West) see grieving as something you do on your own, others see it very differently. As well as taking days, weeks, months or years, the rituals of some societies involve lots of people – sometimes entire tribes or communities.

Psychologist Pittu Laungani describes the process of mourning you might see in an Indian family where the wife has been widowed.* The process is long and expensive, but one detail I particularly like is the practice of more than 50 female friends, neighbours, relatives and well-wishers

* Specifically a high-caste Hindu family in Bombay.

arriving at the widow's house at the same hour every day to mourn together.

Laungani explains that an experienced mourner, sometimes a widow herself, usually takes the lead. Sometimes a 'professional mourner' is hired. They talk about the deceased and try their best to get the members of the family to cry. One sniff, one sob, one shriek leads to another, and another, and soon the room is weeping. He argues that going through this every day has a startling effect on the family's mental well-being and that most reach a point where they can talk about the deceased without holding back.

Although there are no experimental studies on whether the practices of other cultures make death a little easier to handle,* I suspect they do. That's not to say they're ideal – I can imagine there would be times when the last thing you'd want after losing your partner is to have 50 women descend upon you to try to make you cry. But I can also imagine that it might be comforting to have a regular time and space in which to mourn, as well as the knowledge that you're not alone in facing death.

I wondered whether, in cultures where there's little structure or support around death, there's a greater cost. Not only are the bereaved left to deal with their grief alone, but others, the non-bereaved, are free to ardently avoid death, to avoid thinking about it, to avoid feeling it, seeing it, touching it, smelling it, and truly understanding its inevitability – both for those they love, but also for themselves.†

Anthropologist Geoffrey Gorer argues that, just as sex has become more and more mentionable, (natural) death

---

* Or at least, I'm not aware of any such studies.

† Admittedly, in some Western societies people can see their loved ones after death, but it's often not a community ritual and is usually only after they've been cosmetically restored in an attempt to disguise the ravages and decay of death.

has become more and more *un*mentionable; the process of corruption and decay have become disgusting and the ugly facts of death, including grief and mourning, are persistently hidden.* (All the while, violent death has become a fascination.) This, Gorer believes, is the 'pornography' of death, and he argues that if we want to change it we must give back (natural) death its parade and publicity, and readmit grief and mourning into our lives.

You might think that the pornography of death is a good thing – who wants to think about such a gruesome topic if there's nothing we can do about it anyway? Why worry about the things we cannot change? While I can see the appeal of that point of view, I think it misses something crucial: it may only be by really engaging with the fact that your own life is limited, that it will one day end, that you're truly forced to consider how you are living it.

I didn't want a stepmother to my children

'Mine was an arranged marriage. At the time in Sri Lanka, virtually all marriages were. For some reason it surprised me that they worked out. Ours lasted because of my belief that it doesn't matter what ups and downs come in life, you have to stick with it because you are married.' Kumar was a bright-eyed maths teacher in his 60s. He had moved to England in his 30s, shortly before marrying his wife Rani.

'I didn't know anything about my wife. I just had a glimpse of her, that is all. She was good looking, so I nodded to say, "Yes, I will go for it." In the beginning, love wasn't there, but we managed to compromise, and as time

---

* A contributing factor is the hope that many have in modern medicine to cure anyone of anything. As Suzy explained: 'I didn't ever imagine he would die. Especially in this day and age with modern medicine, I just didn't imagine that they couldn't cure someone who was only in his 50s and would cycle about 200km a week. It didn't occur to me. Not until the very end.'

passed, the love grew. Until my wife died, our married life was perfect.

'I lost her 10 years ago. She was in her early 40s and I was about 50, and our three children were very young. My son was barely 12-years-old.

'I didn't tell them she had cancer until the last month. I didn't know how to tell them. I was worried about their current and future mental state, but it came to a point where I had to tell them, I could not hide it. I didn't want them to say, "Why didn't you tell us?" when they grew up one day. I told them that she might leave us and they should be prepared.

'They wanted to know what they could do to keep her alive and had fear of death. My eldest daughter asked, "Why can't they cure it?"* But it was peaceful. There was no outward change in her before she died, and she did not die horribly in pain or suffering, so they did not witness that. We all have to go one day, and it could be tomorrow or 20 days' time; no

---

* Children benefit from being able to ask questions and share their feelings about death – irrespective of how old they are. Not only does this reinforce a sense of trust in their surviving parent, but it also reduces their risk of anxiety. In a Swedish study, teens and young adults who didn't share their feelings about the death of their sibling much or at all were three times more anxious than those who did. It's also particularly important because, unless children know they can talk about these issues if they want to, they can choose to keep quiet, pretend they're doing better than they are or act in a certain way to protect their parents. A UK study of teenagers facing the death of their parent found that many tried to hide their emotions from their parents, and in an Australian study many said they felt guilty and selfish if they tried to talk about their own feelings – but their guilt was alleviated when their parents started the conversation. The danger of not being honest with children, or making them feel that their questions aren't welcome, is that they're free to use their imagination to come up with their own answers – and what they imagine can be far worse than reality. They might, for example, think they're somehow responsible for their parent's illness or death. This sense of responsibility can be even worse in divorce.

one can guarantee our life. In her case, she had to go early and I had to sacrifice my life and comfort my children.

'I was in full-time employment and our three children were under 12. I had never cooked when my wife was alive so when she passed I took over that part. I didn't have a social life and I had to do all the day-to-day things. It was hard and as a man I had to sacrifice a lot to give motherly love. Whatever their mother provided, I provided for them. If the kids said, "Mum did this ..." then I would do the same thing.

'I could have settled with somebody, but I told myself that I needed to do something for my children: I had to show fatherly and motherly love to them because they wouldn't know their mother's love. Love contributes a lot in life. If your parents show you love, that is what you take on board to your future. Material things come secondary, it is the love that's important.

'My sister insisted that I could not live alone with the children – she was one of many people who told me I needed to settle down with someone else. In Asian cultures, there is always back-up, relatives come to stay in the house and help, but in this country everyone is busy and you're expected to just get on with your life. After six months the help started to disappear and no one wanted to know how my heart was after my wife's death. I didn't want material things, but just for people to say, "How are you getting on? How are you coping with the children?"

'I didn't want a stepmother to my children. Suppose I had married somebody and that marriage didn't last – that would have affected them. So, I decided to sacrifice my life until they grew up. That is showing love to your children and not many people are prepared to do that, they will date somebody and off they go. But I decided to focus on guiding them and protecting them. Now they are adults, I can look for love. There is no age limit for that. But love contributes a lot in life and that is why they are flourishing now.'

Before starting this project, I would have thought that talking quite so extensively about death and divorce might become depressing, but I found it strangely clarifying. It also got me thinking, with each interview bringing up different questions in relation to losing a partner. In Kumar's case, it was his philosophy of remarriage that stuck with me. Was he right not to remarry? Is remarrying (with the risk of further loss it brings) bad for children? If you do remarry, is there a better time to do it?

The stress of parenting

It's bad enough having to face your own mortality, suffer the excruciating pain of grief and the vast aloneness that can accompany the death of your partner, as well as deal with the usual barrage of death logistics – but many have to go through all of that at the same time as trying to be a parent to young children or teenagers. In fact, trying to parent after losing your partner can be so difficult that, in a long-term study of 700 widows and widowers, parenting stress caused significantly more distress than grief, anxiety or depression. Just to give you a sense of how much more, visually it would look like this:

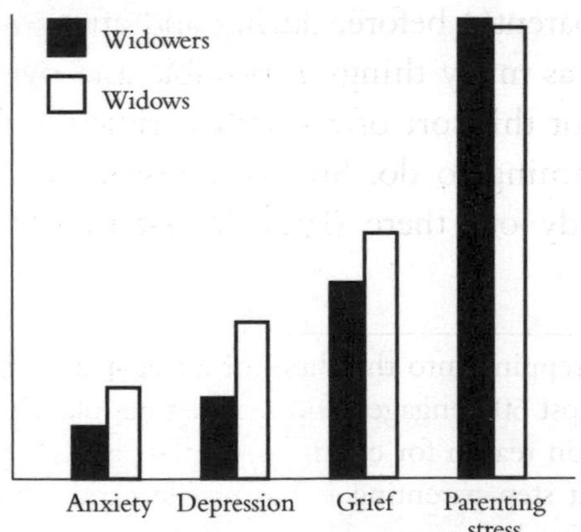

So, I could see why Kumar's sister, among others, advised him to marry again. After all, if he chose wisely, not only would he have companionship, but he would also have someone to help with all the chores, finances and stresses of parenting, and perhaps also a mother figure for his children.*

The remarriage problem

For decades, sociologists and psychologists have debated the impact of remarriage versus single parenthood after divorce or bereavement. Some argue that remarriage is bad for kids, while others believe that children are resilient and can cope with family transitions. Some think that remarriage brings the resources of an extra person to the family, while others argue it can have the opposite effect, diverting energy from the surviving parent away from their children to their new partner (and their kids if they have any). Others point out that growing up with a single parent has its own disadvantages (the main one often being poverty).

Trying to measure the impact of remarriage (and subsequent transitions) on kids isn't easy. As with divorce research, we would ideally have studies that follow children and their parent(s) before, during and after each transition, measuring as many things as possible and over a period of decades. But this sort of research is ridiculously costly and time consuming to do. So many researchers rely on data that's already out there (large long-term studies that are

* Not that stepping into the shoes of a step-parent is easy. In a US study of almost 600 engaged and married people, children were the most common reason for conflict in remarriages. Researchers seem to agree that step-parenting is less problematic when step-parents adopt parenting behaviours slowly, provide support and warmth, and are less forceful around discipline and limit-setting.

made available for researchers to analyse for a variety of purposes) – but this data doesn't ask some of the questions that we might need answers to in order to get the full picture.

For example, a Canadian study in which people were asked to complete questionnaires twice a day, as well as being interviewed by researchers, found that those who remarried and responded to conflict with empathy had much less tension in their relationships. But empathy isn't something that's commonly measured.

Despite the limitations of the evidence, research suggests that remarriage can have a negative impact on kids – although it's not clear why or under what circumstances.* A review of 61 studies looking at kids who had experienced parental death or divorce found that, as well as achieving less academically, those whose surviving or main parent remarried had poorer psychological adjustment than those whose parent didn't. (They also fared worse than children who grew up with two parents who stayed together and didn't die.)† That's not to say the difference is large – it's actually quite small. But it's there.

---

* I say children, but most of the underlying studies looked at teenagers (aged 13–18) – in fact, this made up 80 per cent of the children overall, so the data doesn't accurately represent children across all ages. In fact, it may be much easier for a teenager to experience parental death or divorce than a toddler, simply because the younger a child is when this happens, the more time they have in which to experience more transitions in their life (e.g. remarriage, divorce, etc.).

† I should point out that there is some evidence to the contrary. A large long-term study in Canada did not find a difference between children whose parent died or divorced, and those whose parent died or divorced *and* remarried. In other words, remarriage didn't seem to have an additional negative impact (at least in terms of how likely they were to finish school or not). But perhaps that's because the measure – finishing school or not – doesn't tell us what their grades were like, or how they were doing psychologically.

Timing

If Kumar had wanted to find another partner, would there have been a better or worse time to do so? Unhelpfully, the vast majority of research doesn't address that question, but the little evidence there is suggests that if you are going to get into another relationship, it's best not to do it immediately. (This is particularly relevant for men, as long-term US research found that women are 43 per cent slower than men are to remarry after losing a partner.)*

The Harvard Child Bereavement Study is a long-term US study in which bereaved parents and all their school-aged children were interviewed over the two years following their loss. The study found that children whose parents dated in the first year after bereavement experienced more emotional or behavioural problems, including somatic symptoms, as well as withdrawn and delinquent behaviour. During the second year of bereavement, there were fewer behavioural differences between those whose parents were dating and those who weren't – which suggests that dating is riskier for children's well-being in the immediate aftermath of loss.

It's unclear why dating can have a negative impact. It may not be the dating itself that's the problem, but what drove that parent to date so quickly (for example, trying to fill or escape the gaping chasm of the loss rather than deal with it). Or maybe it's the dominating nature of early lust or romantic love, which might direct their attention away from their children and on to their new partner – just at the time their children need their surviving parent most.

* Interestingly, this study also found that the longer they were married before losing their partner, the slower they were to remarry. This research didn't look at cohabitation (or other forms of relationship) – which is unfortunate given how common it is, and that couples who live together are more likely to break up than those who marry.

If attachment theory is right, making a child feel as loved, safe and secure as possible before, during and particularly immediately after bereavement (or divorce) should help buffer the impact of losing their parent. After all, losing a parent (like losing a partner) means losing an attachment figure – a loss that can have a colossal impact on the attachment style of both parent and child.

Perhaps that's why 59 per cent of children in the Harvard Child Bereavement Study were still concerned about the safety of their surviving parent two years after their loss – a fear of loss that reminded me of anxious attachment – and why the advice of one teenage boy was: after they lose a parent, make sure children feel like nothing will happen to them, make sure they feel *safe*.

Attachment theory echoes the findings of bereavement research, which suggests that one the biggest factors in how children react to death is the relationship they have with their parent who is still alive. In an ideal world, their parent would have good mental health, be consistent with their parenting and discipline, provide them with both continuity and routine, and love, safety, warmth and security.* Perhaps being on a dopamine high would make this already arduous task all the more challenging.†

---

* For example, in a US study of children aged 8–16, researchers found that those whose surviving parent provided warmth and discipline felt less threatened by negative events and better able to cope with them. This was less likely to be the case when their parents had mental health problems.

† And it's pretty difficult in the first place – when you're plummeting into the depths of your own personal grief, your children's needs can be easy to overlook. One father in the Harvard Child Bereavement Study explained that he increased his hours at work to escape – it was only after six months that he realised that his children had needs too. He didn't even think of them before then. Another father said it took him almost a year before he realised that he wasn't the only person who'd lost someone – until then, he hadn't registered that his children had also lost their mother.

Choosing wisely

In order to fully consider the impact of new relationships on children, we also have to factor in how and when they end. Will they break up a few years later? Will the break-up be long, protracted and painful? Will this be followed by another remarriage? Perhaps another break-up? Or will the relationship last for decades until the children have moved out and become fully fledged adults?

The evidence suggests that multiple transitions can have an increasingly negative impact on children. In a large long-term study, researchers analysed the records of all children born or adopted into two-parent families in Manitoba, Canada, in 1984.* They found that children who experienced three or more transitions were significantly less likely to finish school than those whose parents stayed alive and together. Specifically, only 40 per cent of kids who had lived through parental death or divorce, remarriage and then death or divorce, finished school. The percentage of those whose parents stayed together (and alive) was almost double, at 78 per cent.†

This suggests that, as well as thinking about timing, *who* you get into a relationship with is hugely important, because if you get it wrong and decide to walk away, that means yet another transition in your child's life. (And this comes back to the timing point – although I'm not aware of any research on the subject, I strongly suspect that the vortex of acute grief doesn't make for particularly good decision-making.)

---

* The study only followed people who lived continuously in the province until they turned 18 and who were born or adopted at birth into married two-parent households (that's 78 per cent of all children born – a total of 9,906).

† That's a pretty big difference. This remarriage–divorce scenario is more likely for those whose first marriage ended in divorce rather than death (because remarriages after divorce are more likely to end).

I spoke to a number of people who described their parents' poor romantic choices immediately after loss or divorce. Sarah* from the US explained: 'We ended up moving around at least 10 times by the time we were 10-years-old. Mom got married when I was four, so only a year had gone by and we moved into the house with this man and his new family. They divorced and when they split up the only memory that I really have about that is my mom waking us up at two in the morning, throwing all of our clothes into bin bags and saying, "We need to leave right now."

'I don't remember the husband and I don't remember his children. It was a year-long relationship and then my mom remarried when I was five, so maybe 18 months had gone by, and even though that relationship lasted about six years, my sister and I also have the same memory that we don't remember him talking to us at all. Even though he wasn't mean to us or anything, there was never any engagement with the children.

'When my mom got remarried when I was 12, I was there when he proposed to her and my heart sank, I just had anger coursing through my veins because I just felt like, "Here we go again," because by number four, everything is just a pattern now. He moved into the house, he never had any children and didn't really know how to talk to us. I was walking on eggshells because now a strange man was in our house.

'It was so palpable, the awkwardness, and so how did I respond? I started acting like any mischievous teenager does, ditching school, smoking cigarettes, still nobody was paying attention to what I was doing. When I realised that I couldn't get the attention that I wanted and was really upset about her getting married again and having another

*You met Sarah in Chapter 3.

man in the house, I started doing drugs at 14. Although I wasn't conscious of my decision-making at that time, obviously.

'When my mom first found out that I was doing drugs she told me, "Don't fuck this up for me," meaning with her new husband. So that exacerbated my feelings, and of course I responded with anger and decided to go 10 times harder.'

Choose slowly, choose wisely

The research provides us with an insight into trends and averages, but every individual and every relationship is different. Statistically, Sarah was slightly more likely to do poorly at school and drop out than kids who lived with both their parents – but after rebelling, she bucked the trend and did very well academically.* 'Eventually I gave up trying to get attention that way,' she explained, 'I decided to focus on myself, became really pretentious and started reading loads of books. I wanted to go to college and move away as quickly as possible. And that's exactly what I did.'

It's not only children who are unique – so are possible partners and step-parents. If Kumar met someone with a secure attachment style who had no children, spent time with his kids before moving in, was communicative, flexible and keen to be a parent (while being sensitive to the challenges of step-parenting), remarriage would mean something very different to both him and his children than if he partnered with someone who was abusive,

* Although she attributed her anxious attachment style to her upbringing – which goes back to the point that transitions can impact people in very disparate ways, ways that may not be picked up by studies only examining academic achievement, for example.

incommunicative and took little interest in parenting their own children, let alone his.

Parents may only know how a future partner might be with their kids by getting them to spend time together – but, amazingly, this doesn't always happen before people commit. In a UK study of 184 stepfamilies, 47 per cent of children interviewed said they didn't know their step-parent well or at all before they moved in. Even when step-parents and stepkids do spend time together, it might not be that much – research suggests that moving in or remarrying tends to happen very quickly in stepfamilies, leaving children with little opportunity to get to know someone before they become a significant part of their everyday life.

Every relationship, whether parent–child, parent–partner or partner–child, is different. And while remarriage can have a negative impact on kids, much of the time it doesn't – and it's less likely to if parents choose slowly and choose wisely (and if the first time stepfamilies meet is not on the doorstep with a suitcase in hand).

Living life better

'As children, my relationship with my sister was never great and as adults we didn't speak,' said Chloe. She was a receptionist in her early 40s with short, impish hair and high cheekbones. 'She turned up one day after Tim passed away and I told her that she might be an aunty because I was going to try the IVF treatment. She burst into tears and gave me a big hug. She said that on Tim's deathbed she promised him that she would look after me. And she did.

'She gave up smoking, helped give me some of the IVF injections and then moved in just before Florence was born. She stayed for two years – so that's my sister, daughter, me and the dog, all in a one-bedroom flat. We got very close. She would go to work and get home and cook and clean for us. Now we're more like husband and wife than siblings. That is because of Tim dying. I think that we were so stuck in our rut of sniping at each other all the time and I don't think that would ever have changed. I needed something gigantic like this to put things into focus.

'Tim's death gave me a whole new view on the world. I used to be very shy and timid. As a child I had so many regrets of the things that I didn't do because of fear. I still have moments of being scared, but I end up talking myself around. I feel a lot braver now. If I can get through Tim's death and come out with some of my sanity, then I can pretty much get through anything. The worst has happened so you can throw whatever you want at me.

'I remember having a row with Tim's dad. I had always let him bully me before, but for the first time, after Tim died, I stood up for myself. And I decided to carry on with the IVF that we had started before he died. Then I had Florence and I had to bring her up by myself. I never thought I would be capable of that.

'When I'm going through bad things, I try to put things in perspective. I give myself a pep talk, I do that quite regularly … I run through everything that went on and realise that I really don't have anything to worry about. I have got my health and Florence, I have got my sister and family and my pets. It is important to have a moment but not to let it get you down and take over.*

'I realise how precious life is and how quickly it can be taken away. Sometimes I try and say that to people, I try to tell them they haven't got all the time in the world. But even though they know that Tim died at 31, they don't get it. A lot of people I have met think they have

---

* Chloe was speaking 10 years after Tim had died. I suspect that saying this sort of thing to someone whose partner had died in the past few years is a bad idea.

got a full lifetime, they think they are going to die when they are old and grey, and everyone else will die when they are old and grey, but you really don't know what life will throw at you. You have to make the most of the people around you as you don't know what is round the corner for them.'

I cannot tell you how relieved I was to speak to Chloe. I don't know whether it was more for this project, or for myself personally – probably both – but I was desperate to find hope in the face of what can be one of the worst forms of suffering humans can ever experience – losing your partner.

Fear of loss looms so large for some people that they choose (whether consciously or subconsciously) not to love in the first place. Even those who do manage to may feel the deep fear of loss lurking in the crevices of their mind, unknowingly influencing the way they think, feel and behave. Others, meanwhile, expend all their energy pushing it into ever darker cobwebbed corners. So, from the moment I began this project, I was desperate to find a way to make the suffering of loss less petrifying.

The benefits of suffering

No philosopher ever suffered as much as Friedrich Nietzsche. He was misdiagnosed at the time, but he was probably suffering from a retro-orbital meningioma from childhood – a tumour on the brain's surface, which, when left untreated, can grow and grow and grow, squashing the brain into whatever space is left inside the skull. This tumour led to Nietzsche suffering chronic headaches, acute eye pain (eventually leading to blindness in his right eye), visual disturbances, impairments on the left side of the body and, when the tumour ousted his brain's right frontal lobe, dementia.

His headaches were so bad that they would routinely leave him vomiting and bedbound for several days every two or three weeks. By the time he was in his early 50s, his headaches often lasted nine days and he estimated that he lost about a third of the year to them.

This extraordinary amount of suffering had an enormous impact on his philosophy, and Nietzsche came to believe that, without experiencing profound pain, humans lower their capacity for joy. He argued that misery, misfortune and mistakes are as necessary as their opposites – in fact, he thought suffering was so necessary that he went as far as wishing it on the people he cared about.

Over a century later, psychologists have begun to explore the idea that, while deep suffering (including bereavement) is debilitating, many people go on to live more meaningful lives as a result. Much of the research has been in the context of sickness, with studies across Finland, Sweden, India, Canada, China, the US, the UK, Australia and New Zealand finding that people who've suffered from depression to diabetes, chronic fatigue to colorectal cancer, strokes to spinal cord injuries, and HIV to heart disease have changed for the better as a result of their pain. They've developed more empathy, strength and resilience, appreciated the value of time, shifted their priorities (usually valuing their relationships more), and had better relationships as a result.*

The few studies focusing on bereavement have reached similar results. When researchers interviewed nearly 400 people who'd lost someone they loved (17 per cent of which had lost their partner), they found that the vast majority (84 per cent) believed that good had come out of their suffering. A third said that death had strengthened their relationship with family and friends, almost 20 per

*The results vary depending on the study – not all people in all studies across all countries experienced all of these changes. But many did.

cent said they had a greater appreciation for life, and 8 per cent thought they were more patient, understanding, accepting and compassionate.*

In one of the handful of studies focusing on the loss of a partner, psychologist Morton Lieberman found that over a third of the hundreds of widows he interviewed had grown as a result of their loss. They took risks, tried new things, discovered strengths and talents they didn't know they had, and found ways to live in the present rather than for some vague point in the future.

Psychologists Richard Tedeschi and Lawrence Calhoun liken this process of personal growth to experiencing an earthquake. Just as an earthquake has the power to collapse architectural structures that have stood for centuries, loss and suffering have the power to collapse the belief systems that have guided our way of understanding the world, deciding what's meaningful and making decisions. Such seismic change doesn't come easily. It's usually accompanied by profound sorrow – Tedeschi and Calhoun argue that it has to be profound in order to show the cracks by shaking beliefs hard enough.

Just as we can reconstruct buildings after the devastation of earthquakes, we can reconstruct belief systems after loss and trauma. And in the process of doing so, we can create beliefs and ways of being that are more resistant to being shattered. This is what they mean by personal growth.

Before he lost his partner, Andy believed that he was largely in control of his life. But when his partner died in a car crash at 27, Andy's beliefs were shattered. 'Uncertainty

* I was surprised that so few people thought they were more compassionate as a result of their suffering – but this may have something to do with timing. This study approached people about a year after their loss – if they'd been asked years later, the percentage may have been higher.

is difficult, so people try to know everything. And once they feel they do, once they feel they're in control, they feel comfortable,' he said.

'Aaron's death made me realise that I really don't know what's going to happen – and I have found a way to be comfortable with it. Now, I see things in a totally different way – everything somehow falls into place and makes sense. Honestly, I am so happy to be free from the stress of it. Trying to know or control everything is very, very stressful. Now that I've taken a back seat and am mastering not knowing, life is so much easier, so much lighter.'

Not everyone experiences growth. Even among those who do, given the choice, they would still probably choose to have their loved ones back and relinquish all of their acquired wisdom if they could. Growth and meaning don't eradicate pain. In fact, it's precisely because of pain – and specifically the way that people deal with and process it – that leads to personal transformation.*

But I think there's some comfort in the knowledge that people who have lived through the raw, aching emptiness of loss have survived. More than that, many have become better versions of themselves and gone on to live more meaningful lives as a result. And perhaps that's enough to make the inevitability of losing the people we love less daunting and more hopeful.

---

* It's unclear why some people grow as a result of loss and some don't – some suggest that personality comes into it and that people who are more extroverted, reflective or optimistic are more likely to experience growth. But I wonder whether a willingness to really experience the depth of suffering also plays a part. Or perhaps some people in these studies who hadn't grown hadn't yet had enough time to do so. Growth is also less likely when you're just about managing to survive (as is the case in acute grief) compared with when you are actually able to live (in integrated grief).

'I have more empathy for people with hardships now,' explained Suzy from Australia. Her husband of 25 years had died a few years earlier when he was in his early 50s. We spoke in the warmth of the Sydney sun, overlooking the buzz and bustle of the beach below.

'I was a bit superficial and now it doesn't matter any more. I don't worry about my outfits or what people think of me; all that stuff seems so trivial. At one time, I would never leave the house without make-up, but now I don't really care. I don't need to go shopping and buy the latest trendy jacket or whatever, I wear my old stuff and I am fine with that. It is a burden lifted really.

'The idea of getting wrinkles used to freak me out, but looking older or getting older doesn't worry me any more. I had a real thing about it to be honest and started buying all these really expensive cosmetics. Now I just buy supermarket moisturisers and I can't be bothered with all that. I would rather spend the money on helping the kids or a good trip or something.

'I'm less scared of dying. You have no control over any of it: all you can do is live the best life you can and not get freaked out about it. I have relinquished control and it feels like a huge relief. The idea is just to enjoy each day, do what makes me happy and try not to spend time with people that I have got no time for.

'I wish it hadn't happened. And I don't think I will ever stop missing him. But I guess some good has come out of it – Graeme's death has made me spend what I have left of my life better.'

Epilogue

In the process of writing this book, I have come to believe that deep, intimate, meaningful companionate love, whether romantic or not, is one of the greatest joys we can experience as humans – despite the fact that it can, and will, come to an end. Perhaps it's even *because* it will come to an end that it's so precious. After all, isn't it death that makes life quite so valuable?

I have also come to believe that love is a skill that requires knowledge, effort and learning – and I'm in good company; philosophers Nietzsche, Aristotle and Montaigne share similar views. How we develop that skill will depend on our own unique set of defence mechanisms (our suffocating need for closeness, for example, or our denial of unresolved trauma or loss), and how best we can manage them.

As Juan, the 95-year-old farmer from northern Argentina told me at the very start of the project: 'You have to cultivate a relationship, just as you cultivate crops.* If you do not take care of your crops, they will die. It is the same with relationships.' Unless we recognise the role of our own particular defences, we may be more like a diligent farmer who unintentionally poisons his crops instead of cultivating them.

Poison is only part of the story. No matter what our defences, there are ways of thinking and being that can only help with the skill of loving – such as, cultivating compassion, commitment, forgiveness, humility, trust, self-awareness, respect, willpower, thankfulness, courage and kindness. (If you have already nailed those, I'd add trust,

* Juan used the Spanish verb *alimentar* – I've translated it as 'to cultivate', but it can be translated in many ways, including to nourish, foster, nurture or feed. I think relationships require all of these things.

tolerance, sacrifice, patience, acceptance, non-judgement and honesty to the list ...)

Trying to study, learn and practice the skill of loving may sound like an inordinate amount of work. But if you decide that a long-term, committed and meaningful relationship is what you want, why wouldn't you work for it? Why, when education, career and fitness require commitment, self-discipline and hard work, would a long-term, loving relationship require anything less?

The idea of trying to learn how to love may seem nebulous, but it doesn't have to be. Perhaps the key is to start small: choose one thing you think you need to focus on and work on that – whether that's trying to tolerate the irritation of your avoidant attachment system when your partner wants to be close to you, examining whether your expectations are realistic or not, or doing a monkey impression when an argument is escalating out of control.

As well as being responsible for how we love, we're also responsible for who we love. Although passion may be, to some degree, involuntary, dependent on a peculiar cocktail of hormones, immune system compatibility and room temperature, among other things, companionate love is not. This type of love is a decision, and a series of ongoing decisions, including a daily decision to commit and to be open to being changed by someone. (That said, lust isn't completely involuntary and there are plenty of things we can do to rekindle it.)

Just as it falls to us to learn to love, it's also our responsibility to choose someone who will do the same, someone who will value, respect, prioritise, nurture and commit to the relationship and to us. No matter how intoxicating it is, connection, passion or lust cannot overcome the importance, the *necessity*, of learning, understanding and putting in the effort required to love.

If the person you want to be with hasn't developed this skill, and isn't also prepared to put in the hefty psychological work required, as heart-wrenching as it might feel, it may be best to walk away. (Of course, that may be easier said than done. For those in abusive relationships, for example, not only can walking away be traumatic, difficult or dangerous, but it may also be impossible without support.)

Taking responsibility for the role you play in relationships requires courage – huge amounts of it. It's far easier to imagine that the solution lies outside yourself, because that means you can continue to look outward and, in doing so, avoid looking inward. But I firmly believe that it's only in taking responsibility that you can experience the grounding and liberating nature of love.

Acknowledgements

This book wouldn't exist without the hundreds of strangers who gave their time, honesty and support to this project. Despite the mild fear of rejection I felt in approaching them, the moment we began speaking – *really* speaking – all our unknownness melted away and we were simply two human beings working together on a book that we both believed needed to exist. It was a privilege to be able to have so many genuine conversations with people around the world. Thank you.

Thank you also to the academics whose work I relied on in writing this, and in particular those who agreed to be interviewed and/or helped me verify my research: Simon Blackburn, Richard Kraut, Eli Finkel, Peter Backus, Adam Moore, Kidder Smith, Jack Zipes, Miriam Koktvedgaard Zeitzen, Michael Rosen, Julia Hope, Bennett Helm, Simon May, Elaine Hatfield, Kristin Hadfield and Karen Goodall.

Clinical psychologists John Donohue and Damian White – the book is infinitely better for having your insight, gained from years of clinical work and academic study. Damian, thanks especially for your excellent existential brains, and John, thank you for the protracted phone calls about cheery topics, including death, divorce and abuse. You were both fundamental in helping me clarify my thoughts.

Jim Martin – thank you for single-handedly bringing Hawaiian shirts back into fashion and for taking a punt on an unknown author obsessed with relationships. Anna MacDiarmid, thank you for your exceptional edits and your endless support and enthusiasm. Imogen Pelham, thank you for being so effortlessly badass, for seeing the potential in the project and helping me with countless permutations of the proposal. Emily Kearns, your copy-edits were the bomb. Thanks to everyone at Bloomsbury around the world who worked behind the scenes on the project.

Marie Horner – you recorded interviews, listened to countless recordings with me to create the audio sample on my website, and read through the entire book, providing me with invaluable comments. This project is immeasurably better for your wisdom and compassion. Thank you.

Tina Dodds – thank you for transcribing so many interviews over the years and for your long-distance companionship and support. Lana Dettke – thank you for transcribing excerpts from endless journals and books.

Jem Brownlee – if I listed everything you've done to help with this project I'd need another six pages. I'm so appreciative of your brilliant brains. Now write your own book so I can repay the favour.

Kevin Dawson – our conversations about turning the project into an audio series were game-changing and led to me *completely* rewriting the book. Thank you for your outstanding audio and story-telling skills. Thanks also to everyone at Whistledown Productions, including David

Prest and Deborah Dudgeon, for your belief in the project. Andy Staples, Alisdair Hogarth, Verity Wingate, Graham Ross and Cheryl Frances-Hoad – thank you for bringing my thoughts on love to life with music.

Michael Gazzaniga, Morton Liebermann, Michael Rosenfeld, Cindy Hazan, Marjorie Smith, Christine Gross-Loh, Petra at the International Social Science Project, the staff at Japan's Institute of Population and Social Security, and Charlotte and Greg at the Office for National Statistics – thank you for taking the time to respond to my numerous queries.

Deirdre Walker, Jane Park-Weir, John Ball, Joanna Conway and the other brilliant people I worked with at Norton Rose Fulbright – thank you for a phenomenal legal training, which has been instrumental in the process of interviewing, researching, and organising this project.

Thanks to all my friends who took the time to read all or part of the book over the years – SJ Boulos, Pete Farnham, Bea Longmore, Sam O'Leary, Janina Joffe, Peter Smith, Maggie Smith, Alex Faludy, Yoshi Funaki, Jess Moore, Marilyn Boulos, Miranda France, Marina Castledine, Sally Hogbin, Connor Burrowes and Barbara Plaxton – and Jenny Delaney for thinking of the US title of the book. Above and beyond. And a mega thanks to Sapana Agrawal, Tara Button, Pete Farnham, Marie Horner and SJ Boulos for providing 24-hour emotional support.

Jeremy Nagle – for years, you have helped me try endless permutations of search terms and hunt down obscure journals and conference papers that I would not have been able to find myself. Your help was fundamental in keeping the book as inclusive and global as possible – thank you for your extraordinary mind, exceptional attention to detail and generous listening, patience and support. Thanks also to everyone else at the British Library, the research desk (particularly Paul Allchin), librarians, security and cloakroom staff, Roly Keating and all those behind the scenes who keep this invaluable resource working. I love the British Library!

Mum, mum, mum – thank you for always supporting me in whatever I wanted to do, for always letting me know that I was loved, and for being there whenever I needed you. And in terms of this book, thank you for spending so much time looking after B, for helping with late night word-searching and sentence-tinkering, and for always, always having faith in the project – and in me.

B - there's nothing better than a cuddle from you at the end of a long day at my desk. Thank you for (almost) always indulging me.

Ed Smith – you've waited patiently on countless trips around the world while I scuttled off to do interviews, you've listened to me waffle endlessly about love and relationships, you've edited chapters, told more people about the book than I have, and you've cooked, cleaned and looked after B so that I could finish the book in the first year of his life. Thank you. Thank you. THANK YOU. This book wouldn't exist without you. Oh, and given this is a book about relationships, I should probably also thank you for being a fantastic husband ...

Index